TYCHOMANCY

TYCHOMANCY

Inferring Probability from
Causal Structure

MICHAEL STREVENS

Harvard University Press

Cambridge, Massachusetts

London, England

2013

Library of Congress Cataloging-in-Publication Data

Strevens, Michael.
Tychomancy : inferring probability from causal structure / Michael Strevens.
pages cm
Includes bibliographical references and index.
ISBN 978-0-674-07311-1 (alk. paper)
1. Probabilities.   2. Inference.   3. Empiricism.   I. Title.
QC174.85.P76S74 2013
003'.1—dc23   2012042985

For Joy, BBBBB

CONTENTS

Author's Note   *xi*

Physical Intuition   *1*

I

1859

1. The Apriorist   7
2. The Historical Way   *13*
3. The Logical Way   *26*
4. The Cognitive Way   *38*

II

EQUIDYNAMICS

5. Stirring   *53*
6. Shaking   *71*
7. Bouncing   *93*
8. Unifying   *113*

III

BEYOND PHYSICS

9. 1859 Again   *127*

10. Applied Bioequidynamics   *149*

11. Inaccuracy, Error, etc.   *160*

IV

BEFORE AND AFTER

12. The Exogenous Zone   *185*

13. The Elements of Equidynamics   *205*

14. Prehistory and Meta-History   *217*

Notes   *229*

Glossary   *245*

References   *249*

Index   *257*

FIGURES

1.1 Maxwell's velocity distribution   8

2.1 Approach circle for molecular collision   19

3.1 Dodecahedral die   27

4.1 Ball in box with multiple exits   44

5.1 Evolution function for wheel of fortune   56

5.2 Different density, same probability   57

5.3 Nonmicroequiprobabilistic density   58

5.4 Evolution function for two consecutive spins   60

5.5 Evolution function for tossed coin   64

6.1 Bouncing coin landing conditions   72

6.2 Motley wheel of fortune   79

6.3 Motley wheel evolution function   79

6.4 Level one embedding trial   80

6.5 Level two embedding trial   81

7.1 Complex bouncers   109

11.1 Curve-fitting   164

11.2 Three possible error curves   165

12.1 Microequiprobability and rogue variables   187

AUTHOR'S NOTE

A set of rules is presented in this book for inferring the physical probabilities of outcomes from the causal or dynamic properties of the systems that produce them—the rules of what I call *equidynamics*. The probabilities revealed by the equidynamic rules are wide-ranging: they include the probability of getting a 5 on a die roll, the probability distributions found in statistical physics, the probabilities that underlie many prima facie judgments about fitness in evolutionary biology, and more.

Three claims are made about the rules: that they are known, though not fully consciously, to all human beings; that they have played a crucial but unrecognized role in several major scientific innovations; and that they are reliable. The arguments for these claims are, respectively, psychological, historical, and philosophico-mathematical. Psychologists, historians, philosophers, and probability theorists might therefore find something interesting in the following pages, but they will have to find their way through or around some rather foreign-looking material.

Perhaps the most challenging sections are the mathematical arguments. As each of the rules of equidynamics is dredged up from the cognitive depths and displayed in turn, I give some kind of explanation of the rule's ecological validity, that is, its reliability in the environments and contexts in which it is normally put to use. Some of these justifications are reasonably complete, except for formal proofs. Some are mostly complete. For some justifications you are referred elsewhere, although what's elsewhere is itself only partially complete. For some I provide just a hint as to how a justification would proceed. In all cases except the last, the arguments for ecological validity have a certain degree of mathematical complexity. These passages are not easy. Please feel free to skip past them; they are an important part of the complete argument for the theory of equidynamics, but they can be coherently detached from the rest of my case.

It is less easy to skip the history or psychology, but if you are determined, you could read the book as an equidynamics manual with copious illustrations drawn from historical and everyday thought.

Note that the book contains a glossary of technical terms.

My thanks go to Iowa State University for a seed grant to begin thinking about Maxwell and equidynamics back in 1996, to Stanford University for a year of junior leave used partially to draft a long paper about the roots of Maxwell's derivation of the velocity distribution, and to the National Science Foundation for a scholar's grant enabling a year of leave from teaching and administrative responsibilities (2010–2011) during which this book was written. Formal acknowledgment: This material is based upon work supported by the National Science Foundation under Grant No. 0956542.

Thanks also to Michael Friedman, Fred Kronz, and several anonymous NSF reviewers for comments on the research proposal; to audiences at Stanford University and a workshop on probabilistic models of cognitive development held at the Banff International Research Station for comments on talks presenting parts of the project; and to André Ariew, Laura Franklin-Hall, Eric Raidl, and two anonymous Harvard University Press readers for comments on the manuscript.

Well it was an even chance . . .
if my calculations are correct.

TOM STOPPARD,
Rosencrantz and Guildenstern Are Dead

PHYSICAL INTUITION

Do scientists know what they're doing? Not always, not any more than the rest of us. We may do it well, but often enough, we could not say how—how we walk, how we talk, and in many respects, how we think.

Some elements of scientific procedure, such as classical statistical testing or running randomized, controlled experiments, have their essentials made explicit in textbooks or other canonical documents. Some elements, such as the management of various fussy and fragile experimental setups, are typically left tacit, and are communicated from mentor to scientific aspirant through hands-on, supervised learning-by-doing in the lab or field or other practical venue. And some elements involve schemes of inference or other forms of thought that do not seem to be learned at all—or at least, there is no obvious period of apprenticeship in which a student goes from being a novice to a master of the art.

The third class includes a rather heterogeneous cluster of abilities called "physical intuition." One part, perhaps the principal part, of physical intuition is the ability to inspect an everyday situation and, without invoking theoretical principles or performing calculations, to "see" what will happen next; you can, for example, see that putting pressure on the rim of a round, three-legged table midway between any two legs may unbalance the table, but putting pressure on the rim directly over a leg will not (Feynman et al. 2006, 52–53). We humans all have this facility with physical objects, this physical intuition, and we have a corresponding ability to predict and understand without calculation the behavior of organisms, minds, and groups—we have, you might say, biological, psychological, and social intuition.

What we do not have is the ability to see how we see these things: the principles guiding our intuitive physics are opaque to us. We "know more

1

than we can tell" (Polanyi 1961, 467). Or at least, than we can tell right now, since presumably the sciences of the mind are capable of unearthing the architectonics of intuition and thereby explaining our deft handling of furniture and friends.

What does this have to do with real science? Apart, that is, from helping to determine the optimal placement of heavy lab equipment on three-legged tables? Plenty, many writers have thought. Polanyi argued that it is by way of intuition that scientists assess the prima facie plausibility of a hypothesis, and so decide to devote to it the time, energy, and money to formulate, develop, and test it. Rohrlich (1996) adds that great discoveries are made by scientists who use their intuition to distinguish between shortcomings in a theory that may be safely ignored and shortcomings that shred a theory's credibility. Finally, it is received wisdom among those concerned with model-building in science that decisions as to what is essential and what may be omitted from a model are frequently made using a faculty much like physical, or biological, or social intuition.

This story is, as yet, less supported by evidence than by . . . intuition. It is unclear, in particular, that there is a close relation between the abilities used by ordinary people to navigate the everyday physical, biological, and social world, and the abilities used by experienced scientists (let alone great scientists) to assess the prima facie plausibility of hypotheses, models, and theories.

Nevertheless, were Polanyi, Rohrlich, and the other intuitionists to be correct, there would be a largely unexplored dimension along which science and scientific progress might be understood. Theory construction, in that case, even at the highest levels, need not be relegated to the shadowland of "insight" or "genius": on the contrary, some ways in which models and theories are recruited and judged fit for duty will be as amenable to study as any other psychological process. Further, to study such processes, we need not hunt down distinguished scientists and imprison them in MRI machines, since the same patterns of thought are to be found in college sophomores and other willing, plentiful subjects. All that is needed is a bit of history-and-philosophy-of-science glue to hold everything together.

This book is that glue. Or at least, it is the glue for an investigation into one particular aspect of physical intuition: the ability of scientists and ordinary people to look at a physical scenario—or a biological scenario, or a sociological scenario—and to "see" the *physical probabilities* of things and, more generally, to "see" what properties a physical probability distribution over the outcomes of the scenario would have, without experimentation or the gathering of statistical information. I call this ability to infer physical probabilities from physical structure *equidynamics*.

With respect to equidynamics, everything promised above will be found in the following chapters: an elucidation of the rules of thought by which regular people, without the help of statistics, discern physical probabilities in the world; an account of the way in which these everyday equidynamic rules contribute to science by guiding judgments of plausibility and relevance when building models and hypotheses, particularly in physics and biology; and an account of the historical role played by equidynamic thinking in great discoveries.

It is with a very great discovery based almost entirely on equidynamic intuition that I begin . . .

I

1859

1

THE APRIORIST

Why does atmospheric pressure decrease, the higher you go? Why does sodium chloride, but not silver chloride, dissolve in water? Why do complex things break down, fall apart, decay? An important element of the answer to each of these questions is provided by statistical mechanics, a kind of physical thinking that puts a probability distribution over the various possible states of the microscopic constituents of a system—over the positions and velocities of its molecules, for example—and reasons about the system's dynamics by aggregating the microlevel probabilities to determine how the system as a whole will most likely behave.

At the core of statistical mechanics are mathematical postulates that specify probability distributions over the states of fundamental particles, atoms, molecules, and other microlevel building blocks. The discovery of statistical mechanics, this theoretical scaffold that now supports a great part of all physical inquiry, began with the public unveiling of the first attempt to state the exact form of one of these foundational probabilistic postulates.

The date was September 21, 1859. At the meeting of the British Association for the Advancement of Science in Aberdeen, the Scottish physicist James Clerk Maxwell read a paper proposing a probability distribution over the positions and velocities of the molecules of a confined gas at equilibrium—that is, a distribution that the molecules assume in the course of settling down to a steady statistical state (Maxwell 1860).[1]

According to the first part of Maxwell's hypothesis, a molecule in a gas at equilibrium is equally likely to be found anywhere in the space available. The audience would have been neither surprised nor impressed by this suggestion, as it echoed similar assumptions made by previous scientists working on kinetic theory, and more important, conformed to everyday

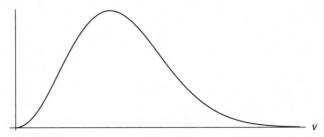

Figure 1.1: Maxwell's proposed distribution over molecular speed

experience (it has never seemed to be harder to get a lungful of air at one end of a room than at the other).[2] What was dramatic about Maxwell's hypothesis was its second part, which went quite beyond the observed properties of the atmosphere and other gases to propose that the components of a gas molecule's velocity are each described by a Gaussian probability distribution, from which it follows that the probability distribution over the magnitude of velocity—the distribution of absolute molecular speed v—has a probability density of the form $av^2e^{-bv^2}$, where a and b are constants determined by the mass of the gas molecules and the temperature of the gas. Figure 1.1 depicts the characteristic form of such a density.

The mathematician and historian of science Clifford Truesdell remarked that Maxwell's derivation of the velocity distribution constitutes "one of the most important passages in physics" (Truesdell 1975, 34). It paid immediate dividends: Maxwell's statistical model predicted that a gas's viscosity is independent of its density, a result that at first seemed dubious to Maxwell but which he then confirmed experimentally. More important, the model as later developed by Maxwell and Ludwig Boltzmann yielded a quantitative explanation of entropy increase in gases, along with the beginnings of an understanding of the microlevel foundation of entropy. Finally, the ideas inherent in the model were applied, again beginning with Maxwell and Boltzmann but also later by J. W. Gibbs and others, to a wide range of physical systems, not limited to gases. Statistical mechanics in its mature form had by then emerged, a vindication of Maxwell's early conviction that "the true logic for this world is the calculus of probabilities" (Garber et al. 1986, 9).

Maxwell's proposed velocity distribution for gas molecules, the first tentative step toward modern statistical physics, need not have been accurate to have had the influence that it did. It would have been enough to point the way to quantitative thinking about physical probabilities;

further, some results about gases, including the viscosity result, do not depend on the exact form of the distribution. Nevertheless, the velocity distribution was quite correct.

Direct confirmation of the distribution's correctness did not arrive until long after Maxwell's death when, in 1920, Otto Stern applied his "method of molecular rays" to the problem. Stern's apparatus allows a brief "puff" to escape from a gas of heated metal atoms; the velocities of the molecules in this burst are then measured by rapidly drawing a sheet of glass or something similar past the site of the escape. When they hit the sheet the molecules come to rest, forming a deposit. Because the fastest molecules in the burst reach the sheet first and the slowest molecules last, and because the sheet is moving, the fast molecules will deposit themselves at the head of the sheet, the slow molecules at its tail, and molecules of intermediate speeds in between at positions proportional to their speed. The distribution of molecular velocity will therefore be laid out along the sheet in the form of a metallic film. Stern showed that the density of the film mirrored the density of Maxwell's velocity distribution (Stern 1920a, 1920b). He considered this as much a test of his new apparatus as a test of Maxwell's hypothesis, so secure was the theoretical status of the Maxwell distribution by this time.

If Maxwell did not measure the velocities of molecules himself, how did he arrive at his probabilistic postulate? Did he work backward from the known behavior of gases, finding the only distribution that would predict some known quirk of gaseous phenomena? Not at all. Maxwell's derivation of the velocity distribution rather appears to be entirely a priori. "One of the most important passages in physics"—the seed that spawned all of statistical mechanics—was a product of the intellect alone, empirically untempered. Could this semblance of aprioricity be the real thing? Could Maxwell have directly intuited, in some quasi-Kantian fashion, the precise shape of the actual distribution of molecular velocity? Did he make a lucky guess? Or was something else going on? This book aims to answer these questions and to generalize the answer to similar probabilistic inferences in other domains of scientific inquiry.

Maxwell's derivation of the velocity distribution is, on the surface, short and simple. Grant that gases are made up of myriad particles, microscopic and fast-moving. (This assumption, still controversial in 1859, was of course empirical; what was a priori about Maxwell's reasoning, if anything, was the statistical component of his model of a gas.) What can be supposed about the velocities of these molecules at equilibrium?

Three things, Maxwell suggests. First, the probability of a molecule having a certain velocity will be a function of the magnitude of that velocity alone; or, in other words, molecules with a given speed are equally likely to be traveling in any direction. Second, the probability distributions over each of the three Cartesian components of a molecule's velocity—its velocities in the x, y, and z directions—are the same. Third, the three Cartesian components of velocity are independent of one another. Learning a molecule's velocity in the x direction will, for example, tell you nothing about its velocity in the y direction.

These assumptions put what might look like rather weak constraints on the molecular velocity distribution. In fact, they are anything but weak: there is only one distribution that satisfies them all. Maxwell concludes that the distribution of molecular velocity must assume this form. (The mathematical details of the derivation are set out in the appendix at the end of this chapter.)

Naturally, then, the reader asks: what is the basis of the three assumptions, so strong as to uniquely determine the velocity distribution, and correctly so? Maxwell makes the first assumption, that the probability of a velocity depends only on its magnitude, on the grounds that "the directions of the coordinates are perfectly arbitrary" (Maxwell 1860, 381), that is, that for the purpose of providing a Cartesian representation of molecular velocity, you could choose any three mutually orthogonal lines to be the x, y, and z axes. He does not provide an explicit justification for the second assumption, but he has no need of one, because it follows from the other two (Truesdell 1975, 37). The rationale for the third assumption, the independence of a molecular velocity's three Cartesian components, is as follows: "The existence of the velocity x does not in any way affect that of the velocities y or z, since these are all at right angles to each other and independent" (p. 380).

The premises from which Maxwell deduces his velocity distribution are two, then: first the conventionality of, and second the mathematical independence of, the three axes that contribute to the Cartesian representation of velocity. Each of these is, if anything is, a conceptual truth. Maxwell thus appears to have derived his knowledge of the actual distribution of molecular velocities on entirely a priori grounds: apart from the empirical presuppositions of the question—that there are molecules and that they have a range of velocities—the answer seems to rest on abstract philosophical or mathematical truths alone, floating free of the particular, contingent physics of our world and readily apparent to the armchair-bound intellect.

Had an eminent mind claimed physical knowledge based on pure reflection in the seventeenth century, you would not be surprised; nor would you expect their claims, however historically and intellectually interesting they might be, to correspond to physical reality. But it was the nineteenth century. And Maxwell was right.

This ostensible vindication of untrammeled apriorism demands a closer look. What was going on under the surface of Maxwell's text? How might considerations other than purely conceptual observations about the Cartesian system of representation have guided Maxwell's argumentation, or given him the confidence to present the conclusions that he did?

There are three places to search for clues, to which the next three chapters correspond. First there are what I will call, in the broadest sense, the historical facts, by which I mean primarily the scientific writings of Maxwell and his predecessors. Second, there are what I will call the logical facts, by which I mean the facts as to what forms of reasoning, a priori or otherwise, give reliable or at least warranted information about physical probability distributions. Third, there are the psychological facts, the facts about what forms of reasoning to conclusions about probability distributions—if any—are typically found in human thought.

Appendix: Maxwell's Derivation of the Velocity Distribution

Maxwell supposes that the same density $f(\cdot)$ represents the probability distribution over each of the three Cartesian components of a molecule's velocity. Further, the components are stochastically independent, so the probability density over velocity as a whole is a simple function of the densities over the components:

$$F(x,y,z) = f(x)f(y)f(z)$$

At the same time, Maxwell assumes that the probability distribution over velocity as a whole depends only on velocity magnitude, and so may be represented as a function of magnitude squared alone:

$$F(x,y,z) = G(x^2 + y^2 + z^2)$$

for some choice of G. The form of the distribution, then, is such that

$$G(x^2 + y^2 + z^2) = f(x)f(y)f(z)$$

for some choice of f and G.

Surprisingly, only one choice of f satisfies this constraint:

$$f(x) = ae^{bx^2}$$

As with any probability density, the area under this function must sum to one, from which it follows that the coefficient of the exponent is negative. The component density f is therefore a Gaussian distribution.

2

THE HISTORICAL WAY

Why did Maxwell find his derivation of the velocity distribution convincing or, at least, plausible enough to present for public consumption? Why did a significant portion of his public—the physicists of his day—regard it in turn as a promising basis for work on the behavior of gases?[1] And how did he get it right—what was it about his train of thought, if anything, that explains how he hit upon a distribution that not only commanded respect from his peers but accurately reflected reality? The same distribution made manifest in Otto Stern's molecular deposits sixty years later?

Under the heading of history, I look for answers to these questions not only in the broad currents of nineteenth-century scientific thinking, but also in Maxwell's own writings, and in particular—and, as it will turn out, most importantly—in the other parts of the paper in which he derives his velocity distribution.

2.1 Proposition IV

Begin with "history" in the narrowest sense, that is, with Maxwell's overt reasons for postulating the velocity distribution. The derivation occupies just the few paragraphs that make up proposition IV of Maxwell's 1859 paper on gases (Maxwell 1860). Put on hermeneutic blinders and imagine for a moment that nothing else exists, that proposition IV contains everything relevant to Maxwell's reasoning. How convincing, and how truth-conducive, is the derivation?

It is remarkably weak.

The first premise of the derivation, that the probability of a velocity depends only on its magnitude, is equivalent to the assumption that the distribution over velocity is spherically symmetrical: it looks the same from all directions, or in other words, however you rotate it.

The assumption has a certain plausibility. But Maxwell's reason for advancing it is unconvincing. He observes that the directions of the Cartesian axes used to represent velocity are arbitrary, something that, being a conceptual truth about the nature of Cartesian representation, is true for any probability distribution over velocity, or indeed, over any physical quantity represented in three-dimensional space. But it is obvious that not all such distributions will have spherical symmetry. The distribution over the positions of gas molecules in a significant gravitational field, for example, tails off with height (that is, distance from the gravitational attractor).[2]

Thinking about this and other cases, you should see that the symmetry assumption is entirely unmotivated by the arbitrariness of the coordinate system. Something else must give the assumption its credibility, presumably something that is present in a normal enclosed gas but not in a gas subject to a gravitational field or similar force—something physical, then.

So facts about physical symmetries or asymmetries play a role in persuading us to accept Maxwell's first premise. What facts? I can hardly wait to answer this question, but for now it is postponed: this is the history chapter, in which I restrict myself to justifications offered on the record, or at least hinted at, by Maxwell and his contemporaries.

The second Maxwellian premise, the stochastic independence of the Cartesian components of velocity, has less prima facie plausibility than the first. Maxwell's justification of the premise does not improve the epistemic situation: he offers nothing more than the observation that the Cartesian components are independent in the mathematical sense (a consequence of the orthogonality of the axes). This is better than wordplay, since mathematical independence is (more or less) a necessary condition for stochastic independence, but it is not much better, being far from a sufficient condition. Maxwell himself later conceded that his reasoning was at this point "precarious" (Maxwell 1867, 43).

Yet even Maxwell's own later misgivings cannot annul the fact that his reasoning, however logically insecure, led him to true conclusions with momentous consequences—momentous in part because a few of his more extraordinary readers, not least Boltzmann, thought that he had gotten something important right.

A close examination of Maxwell's official reasons for making the assumptions he does, then, aggravates rather than alleviates the puzzle.

Maxwell not only made a great empirical discovery by way of an entirely a priori argument—it was a bad a priori argument.

Again, it should be asked: did he just get lucky? Did he, in the course of playing with statistical models constructed more with the aim of simplicity than veracity, stumble on the truth? But then how did he recognize it for the truth, or at least, for a serious possibility? How did his peers do the same?

2.2 The Zeitgeist

Maxwell's derivation was not original—not in its mathematical skeleton. Partway through a long 1850 review of a book by Adolphe Quetelet, John Herschel used a similar argument to justify the supposition that errors in a wide variety of scientific measurements assume a Gaussian distribution, so vindicating the use of the "method of least squares" to deal with measurement error (Herschel 1850; see also section 11.1 in this volume).

Herschel reasons as follows. He compares the process of making astronomical measurements—of pointing a telescope at a celestial object to measure, say, its position from day to day—with a marksman firing a rifle at a target or a scientist repeatedly dropping a ball from a great height attempting to hit a specified point on the ground below. Each of these processes will be subject to error, and in each case three assumptions can, according to Herschel, be made about the error:

1. There is a probability distribution over the possible errors that takes the same mathematical form for every such process; that is, there is a single probability distribution describing the errors in rifle-firing, telescope-pointing, and ball-dropping (though presumably the parameters of the distribution vary from case to case).
2. The probability of a given error depends only on its magnitude; thus, an error of a given magnitude is equally likely to be in any direction, or equivalently, the distribution over errors in the two-dimensional target space has circular symmetry.
3. The Cartesian components of any given error are independent.

From the second and third premises, mathematically equivalent to the premises of Maxwell's derivation (though Herschel's measurement errors are represented in two-dimensional space, whereas molecular velocities are represented in three dimensions), Herschel concludes that the distributions over the Cartesian x and y components of error are Gaussian, and consequently that the distribution over the magnitude m of error has the form ame^{-bm^2}.

The principal difference between Herschel's and Maxwell's derivations, aside from the subject matter, is in the grounds that they give for their assumptions. Maxwell's grounds are conceptual truths about Cartesian representation. Herschel adds to the mix considerations of ignorance. First, he argues that the probability distribution over errors will be the same regardless of the source of the error, because we are equally ignorant of the causes in all cases; for example, because we know nothing to distinguish the causes of error in marksmanship and telescopy (or at least, we know nothing distinctive about the statistical distribution of the causes), we should suppose that the same probability distribution describes both. Second, he gives the same reason—"our state of *complete* ignorance of the causes of error, and their mode of action" (p. 20)—for making the assumption of circular symmetry, that is, for supposing that the probability distribution depends only on the magnitude of the error. (Precisely why our ignorance justifies the assumption that the probability of an error may depend on its magnitude but not on its direction, rather than, say, vice versa, is unclear, though of course there are simple nonepistemic reasons why the probability must eventually taper off with increasing magnitude.) Ignorance does not come into Herschel's justification of the component independence assumption, which appears to be similar to the justification offered by Maxwell.[3]

Despite the central role played by ignorance in grounding his Gaussian error distribution, Herschel expects actual frequencies of errors to conform to the distribution:

> Hence this remarkable conclusion, viz. that if an exceedingly large number of measures, weights, or other numerical determinations of any constant magnitude, be taken,—supposing no bias, or any cause of error acting preferably in one direction, to exist— . . . the results will be found to group themselves . . . according to one invariable law of numbers [i.e., the law of errors] (Herschel 1850, 20-1).[4]

For this attempt to create statistical knowledge from an epistemic void, Herschel was roundly criticized by R. L. Ellis (1850), who made many of the same objections you would hear from a philosopher of science today (see section 3.2).

Was Maxwell influenced by Herschel? Historians of physics agree that he was likely familiar with Herschel's argument by the time he derived the velocity distribution (Garber 1972). He may well have taken the idea from Herschel, then, while purging it of its subjectivist elements, that is, of its reliance on considerations of ignorance. It is possible also that Herschel's immense standing in the world of British science gave Maxwell some confidence in the force of the argument. (It is unclear whether he was familiar with Ellis's and others' objections to Herschel.)[5]

However, it would surely have been reckless for Maxwell to publish his theory of gases on these grounds alone. His decision to delete the subjectivist aspect of Herschel's argument suggests that he did not find it entirely satisfactory as it stood. More important, whereas Herschel was explaining a statistical pattern for which empirical evidence was accumulating rapidly—the Gaussian distribution of measurement error—Maxwell was predicting a statistical pattern for which there was no empirical evidence whatsoever.

Further, that a star-struck Maxwell swallowed Herschel's argument without complaint hardly explains the extraordinary fact that his subsequent reasoning led him directly to the velocity distribution's true form. It would be a mistake to assume that Maxwell's success could only be explained by his using an infallible or entirely rational method, but it would be equally a mistake to attribute his success to pure chance, holding that it was simply Maxwell's good fortune that fashionable ideas about the distribution of errors happened to point to the right probability distribution for gaseous molecular velocities. Or rather, it would be a mistake to settle for the luck hypothesis without first looking harder for alternatives.

The sociocultural sources of Maxwell's 1859 derivation surely extend beyond Herschel's review, but how far and in what directions we may never know; certainly, historians have uncovered little else about the possible influences on the argument's specific form. Let me therefore return to textual analysis of Maxwell's 1859 derivation of the velocity distribution.

2.3 The Road to Proposition IV

Maxwell's apparently a priori discovery of the velocity distribution appears in the 1859 paper's proposition IV. If this passage deduces from first principles the conceptual foundation for everything that is to follow, what is in propositions I through III?

Proposition IV derives the form of the molecular velocity distribution for a gas that has settled down to a statistically steady state. Propositions I, II, and III purport to establish the existence of such a steady state, that is, they purport to establish that over time, the distribution of velocities in a gas will converge to a unique fixed distribution—that the velocity distribution has a global, stable, equilibrium. The first three propositions play a similar role, then, to Herschel's argument that, because of our ignorance of the causes of error, there is a unique distribution over measurement errors of all types.

Maxwell's argument for the equilibrium runs as follows. Proposition I lays out the physics of collisions between perfectly elastic hard spheres—the physics of idealized billiard balls. (Maxwell has already remarked in the paper's introduction that his conclusions will apply equally to particles that do not collide but that repel one another by way of strong short-range forces.) As yet, no statistical considerations are introduced.

Proposition II aims to calculate the probability distribution over the rebound angle resulting from such a collision. Fix the frame of reference so that one sphere is not moving. Then the other sphere, if there is to be a collision, must be moving toward the fixed sphere. More specifically, the moving sphere's center must pass through a circle orthogonal to its direction of motion, a circle whose center lies on a line parallel to the direction of motion emanating from the center of the fixed sphere, and whose radius is equal to the sum of the two spheres' radii, as shown in figure 2.1. Call this the *approach circle*. (The notional approach circle may lie at any point between the two spheres; it does not matter where.)

Maxwell now introduces a statistical postulate: he assumes that, conditional on the occurrence of a collision, the moving sphere's center is equally likely to have passed through any point of the approach circle. Call the point where the sphere's center passes through the circle the sphere's *approach point;* then Maxwell's assumption is that any approach point within the circle is equally probable, or in other words, that the probability distribution over the approach point, conditional on a collision's taking place, is uniform. (Like Maxwell, I will not stop to ask in virtue of what facts there could be such a probability distribution in a world whose laws were, as Maxwell believed, deterministic.) From the uniformity assumption and the physics of collisions, Maxwell shows that

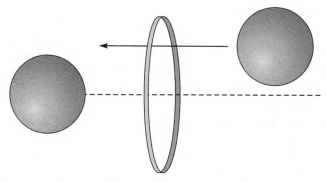

Figure 2.1: The approach circle: the two spheres collide just in case the center of the moving sphere (right) passes through the approach circle

the moving sphere is equally likely to rebound in any direction, that is, that its velocity after the collision is equally likely to be in any direction.[6]

Proposition III brings together the deterministic physics of collisions with the rebound angle equiprobability derived in proposition II; it states that the velocity of a sphere after a collision with another sphere is the sum of the velocity of the spheres' center of mass and a velocity determined by the impact itself, which will be with equal probability in any direction, relative to the center of mass (since the rebound angle in the center-of-mass frame of reference will be the same as in proposition II's fixed-sphere frame). In effect, the collision subjects the velocity of the sphere to a random adjustment.

Maxwell then concludes:

> If a great many equal spherical particles were in motion in a perfectly elastic vessel, collisions would take place among the particles, and their velocities would be altered at every collision; so that after a certain time the *vis viva* [kinetic energy] will be divided among the particles according to some regular law, the average number of particles whose velocity lies between certain limits being ascertainable, though the velocity of each particle changes at every collision (Maxwell 1860, 380).

In other words: the cumulative effect of the random adjustments of velocity due to manifold collisions will converge on a single distribution (the "regular law"). Maxwell's proposition IV follows, opening with the stated aim of finding "the [velocity distribution] after a great number of collisions among a great number of equal particles" (where "equal" means "equally massive"). At this point, Maxwell presents the apparently a priori derivation of the velocity distribution.

Why did Maxwell think that the aggregate effect of many random velocity adjustments would impose a fixed and lasting distribution over velocity? He does not say, but the conclusion hardly follows from its grammatical antecedent, that the molecules' velocities "would be altered at every collision."

You might guess that Maxwell is reasoning thus. The velocity of a particle, after a large number of collisions, will be the sum of its initial velocity and the changes in velocity effected by the collisions. After many collisions, the initial velocity will comprise a negligible portion of this sum—it will have been "washed out." Thus a molecule's velocity, after a sufficiently long period of time, will be for all practical purposes determined by the distribution of the changes in velocity effected by collisions. If there is a fixed distribution over these changes that does not itself depend on the molecules' initial conditions, then there will be, "after a certain time," a fixed distribution over the velocities themselves.

In supposing the existence of the latter fixed distribution—the distribution over changes in velocity—Maxwell might have taken some comfort from de Moivre's theorem and Laplace's more general central limit theorem showing that iterated random fluctuations of the right sort converge, if independent, to a Gaussian distribution. (To secure independence, he would have to assume that the approach points for any two collisions are independent of each other, a postulate that seems as plausible as the uniformity assumption itself.) But if Maxwell saw his conclusion as hinging on these powerful mathematical results, it is curious that he does not think to say so.

I will argue eventually (see section 8.4) that Maxwell's reasoning does turn on "washing out" of the sort described above, and I will explain why many major premises of the washing-out argument are nevertheless missing from the text.

Until then, let me rest with a provisional conclusion. Propositions i through iii have an important role to play in Maxwell's derivation of the velocity distribution; their function is to establish that there is a unique equilibrium distribution to derive. There is clear textual evidence for this claim: as noted above, after the final sentence of proposition iii declares that the velocity distribution of "a great many . . . particles" settles down as a result of collisions to an equilibrium "after a certain time," the first sentence of proposition IV declares explicitly the intent to derive the form of the velocity distribution resulting from "a great number of collisions among a great number of . . . particles." Maxwell's later summary of the first part of the paper confirms in passing the role of the approach to equilibrium (1860, 392).[7] Though the argument for the existence of the equilibrium distribution is perhaps somewhat obscure, there can be no question that Maxwell attempted such an argument and that he regarded it as an integral part of the derivation of the velocity distribution.

If my provisional conclusion is correct, then Maxwell's derivation of the distribution is not, after all, a priori, for one and a half reasons. First, the derivation of rebound-angle equiprobability requires assumptions about the physics of intermolecular collisions. Although Maxwell's assumptions were based on surmise rather than empirical testing, he of course made no claim to discern the physics a priori. Second—and this is the half-reason—the derivation requires assumptions about the probability distribution over approach points. Unlike the probabilistic assumptions in proposition IV, Maxwell does not base these on conceptual truths about Cartesian representation. Nor does he give them any other rationale. He simply asserts in proposition II that "within [the approach] circle every position is equally probable," as if it were indisputable. Indisputable because a priori? Maxwell does not say.

I now want to go back to reexamine Maxwell's grounds for making the statistical assumptions that serve as the premises of his Herschelian argument in proposition IV, namely, the equiprobability of velocities of equal magnitude and the independence of the three Cartesian components of velocity. The probabilistic reasoning of propositions I through III, I will propose, provides an alternative and more plausible grounding for the proposition IV assumptions than Maxwell's official grounding in the formal properties of Cartesian representation. Although Maxwell did not make this alternative grounding explicit, it played a role, I conjecture, in convincing him and some of his readers of the validity of the proposition IV assumptions, and equally importantly, it helps to explain how Maxwell so easily hit upon the truth.

Begin with the assumption that a gas molecule with a given speed is equally likely to be traveling in any direction—that is, the assumption that the velocity distribution has spherical symmetry. There is a justification for this assumption that closely parallels Maxwell's putative rationale, suggested above, for the existence of a fixed long-run velocity distribution. It goes as follows. From propositions II and III, the velocity distribution over changes in a molecule's velocity is spherically symmetric. It follows, if these changes are stochastically independent, that the distribution over a series of such changes is also spherically symmetric. Since the velocity distribution is determined, in the long run, by this spherically symmetric distribution—initial conditions are "washed out"—it too must be spherically symmetric.

The argument is not quite correct, however: the distribution over changes in velocity relative to a colliding pair's center of mass is spherical, but different collisions have different centers of mass. Before aggregating the changes, then, they ought to be converted into a common frame of reference; the obvious choice would be the rest frame with respect to which velocities are represented by the velocity distribution. This is not straightforward, as subsequent attempts to provide more rigorous grounds for kinetic theory have shown. The task is not impossible, but it is probably not reasonable to suppose that it was accomplished by Maxwell.

Another, surer line of thought was available to Maxwell, however. He had concluded, at the end of proposition III, that the distribution of molecular velocities converges over time to a unique distribution—a global, stable, equilibrium. Once established, this distribution, being a stable equilibrium, remains the same; in particular, it does not have any tendency to change in the short term, minor fluctuations aside. Therefore, a distribution of velocities (or positions) that does change in the short term cannot be the equilibrium distribution, whereas a distribution that does not change must be the equilibrium distribution, since there is only one equilibrium. By examining short-term trends in the behavior of different distributions, then, it is possible to learn the properties of the equilibrium distribution.

Suppose that a gas's velocities are distributed so that the first Herschelian premise is false: among molecules traveling equally fast, some directions of travel are more probable than others. Because of rebound angle equiprobability, intermolecular collisions will very likely soon begin to undo this bias: some more probable directions of travel will become less probable than before, while some less probable directions become more probable than before. No biased distribution is stable; therefore, the equilibrium distribution cannot be a biased distribution. It must be a distribution in which a molecule traveling at a given speed is equally likely to be traveling in any direction.[8]

Or you might reason the other way. A distribution in which velocities of equal magnitude are equiprobable is stable, because no particular directional bias is favored in the short term by the stochastic dynamics of collisions established in propositions I to III. Therefore, such a distribution must be the equilibrium distribution.

Note that even the conjunction of the above two suppositions about short-term stability, amounting to the proposition that the only stable velocity distribution is one in which direction given magnitude is equiprobable, is insufficient to establish the existence of a global equilibrium: it is consistent with the possibility that for many initial states, a gas's velocity might never equilibrate, that is, might never settle down to a single, fixed distribution.

The same kind of reasoning may be used to justify the second Herschelian premise, that the Cartesian components of velocity are stochastically independent. In this case, the relevant short-term probabilistic trend breaks down correlations among the Cartesian components of a molecule's velocity. As a result of this trend, no distribution with correlated components has short-term stability; therefore, no such distribution is the equilibrium distribution, and so the components in the equilibrium distribution are independent. The trend to Cartesian dissociation is, I think, less evident than the trend dissociating direction and magnitude of speed; further, Maxwell's investigations in propositions I to III do less to establish its existence than they do to establish the existence of direction/magnitude dissociation. But still, such a trend can be faintly discerned by the unaided human intellect.

What evidence is there that Maxwell's confidence in the Herschelian premises was based on this line of reasoning, this relationship between long-term equilibrium and short-term dynamics? In proposition IV, no evidence whatsoever. But later in the 1859 paper, similar conclusions are based on argumentation of precisely this form.

In proposition XXIII, for example, Maxwell sets out to show that the kinetic energy of a gas composed of nonspherical molecules will become equally distributed among the three Cartesian components of (translational) velocity and the three Cartesian components of angular velocity "after many collisions among many bodies." Invoking the velocity distribution as a premise, he demonstrates that only an equal distribution of energy between the translational and angular velocities is stable in the short term, and concludes that such a distribution will be "the final state" of any such gas (Maxwell 1860, 408–409). Maxwell therefore reasons, in this passage, from facts about short-term stability and instability to the properties of the equilibrium distribution, and does so by way of an argument that presumes the existence of a unique, global, stable equilibrium—his "final state." Without this assumption the argument fails, as in itself it gives no reason to suppose that the state of equal energy distribution is stable, let alone that a system starting out in an arbitrary nonequilibrium state will eventually find its way to the equilibrium.

With a little more hermeneutic effort, which I leave as an exercise to the reader, the same kind of reasoning can be discerned in proposition VI; what is more, Maxwell later used a similar approach to provide a new foundation for the velocity distribution in his "second kinetic theory" (Maxwell 1867).

It would be peculiar if Maxwell, having developed a stochastic dynamics for molecular collisions and having used it to argue for the equilibration of the velocity distribution in propositions I to III, and then having used both the short-term dynamics and the putative fact of global equilibration to derive properties of the equilibrium distribution in other parts of his paper (propositions VI and XXIII), entirely ignored the relevance of the same kind of argument for his Herschelian posits in proposition IV.

Although the official justification of the posits makes no mention of such considerations, then, there lie in Maxwell's text abundant logical, physical, and probabilistic materials for the construction of an alternative, unofficial justification for the posits that is considerably more convincing. It is no great leap to suppose that the force of this unofficial argument, however dimly perceived, played some role in encouraging Maxwell to continue his researches and to publish his paper, and also perhaps in encouraging his more perceptive readers to take Maxwell's statistical model seriously.

How did Maxwell's paper come to contain two rival arguments? I suspect that the paper was shaped originally around what I am calling the unofficial argument, that is, the reasoning based on the equilibrating effect of the accumulation of many independent adjustments of molecular motion described in proposition III. Maxwell saw that the argument was not mathematically complete and, dissatisfied, at some point inserted the alternative a priori argument for the Herschelian posits, severing proposition IV's connection to the preceding three propositions and introducing an official argument whose mathematics is as ineluctable as its logic is inscrutable. If this is correct then the official, a priori argument played no role whatsoever in Maxwell's discovery of the velocity distribution; it is merely a post hoc philosophical papering over of the flawed but vastly more fruitful unofficial argument that guided Maxwell to the insights on which statistical physics was built.

Putting these speculations aside, it is in any case the unofficial argument, I have suggested, that explains both the rhetorical power and the truth-conduciveness of Maxwell's reasoning. But this is possible only if the lacunae in the argument for the existence of a global equilibrium were somehow logically and psychologically patched and only if the probabilistic posits upon which the equilibration argument is itself based, the uniform distribution of and the stochastic independence of the pre-collision approach points, were themselves plausible and close enough to true.

You might think that the uniformity and independence posits alone sink the story. What reason is there to suppose that they are correct, let alone transparently so to nineteenth-century readers? We have gone from probabilistic postulates with a questionable basis—the postulates of proposition IV with their foundation in conceptual truths about Cartesian representation—to probabilistic postulates with no visible basis at all. Is this an improvement?

History can only get you so far. So much reasoning goes unrecorded—neither permanently recorded in the annals of science, nor even ephemerally recorded in the mind's eye. So much reasoning is simply unconscious. Other methods are needed to dig it out.

3

THE LOGICAL WAY

The scientific power of Maxwell's first paper on statistical physics sprang in large part, I have proposed, from an unofficial, alternative argument in which the mysterious premises of proposition IV, the equiprobability of velocity's direction given its magnitude and the independence of its Cartesian components, function as intermediate steps rather than as foundations. The real foundations are specified in propositions I through III. They are of two kinds: dynamic facts about the physics of collisions, and further probabilistic posits.

The more salient of the new probabilistic posits is the assumption that a molecule is in some sense equally likely, en route to a collision, to pass through any point on the approach circle (section 2.3). This is presumably a special case of a more general assumption: the probability distribution over molecular position is approximately uniform over any very small area.

I have also tentatively attributed to Maxwell a second probabilistic posit, concerning the stochastic independence of the approach points of any two collisions. Maxwell does not make this assumption explicit, but it or something similar must be attributed to him to make sense of his reasoning about the cumulative effects of many collisions. I will not make any further specific claims about the form of Maxwell's independence posit until chapters 7 and 8, but bear in mind its existence.

Call these two posits together the assumption of the *microequiprobability* of position. (Microequiprobability incorporates some sort of stochastic independence, then, even though independence makes no contribution to its name. Characterizations of new technical terms introduced in this book may be found in the glossary.) What advantage is gained by basing the derivation of the velocity distribution on the microequiprobability of position rather than the proposition IV posits?

Microequiprobability has a kind of intuitive rightness that the proposition IV posits lack. Simply to assume the independence of the Cartesian components of velocity is rebarbative; to assume it on the grounds that the components are mathematically independent is if anything even more so. (Better, sometimes, not even to try to explain yourself.) To assume that positions are microequiprobable seems, by contrast, rather reasonable.

Why? What is the source of our epistemic comfort? Is it emotional, rhetorical, mathematical, physical? Or philosophical? Logical?

3.1 Doors and Dice

You must choose between three doors. Behind one is a hungry saber-toothed tiger. Behind another is a lump of coal. Behind the third is a well-upholstered endowed chair. That is all you know. Is there any reason to prefer one door over the others? Apparently not. You should consider the tiger equally likely to lurk behind any of the three doors; likewise the chair.

You roll an unfamiliar looking die, in the shape of a dodecahedron (figure 3.1). Its faces are numbered 1 through 12. In a series of rolls, is

Figure 3.1: A dodecahedral die

there any reason to expect one face—say, the 5—to turn up more often than the others? Apparently not. You should consider a roll of 5 to be just as likely as any other roll.

In both cases, you note some symmetry in your situation—the three identical doors, the twelve identical faces—and from that symmetry you derive a probability distribution that reflects the symmetry, assigning equal probabilities to relevantly similar outcomes. This is the probability distribution that seems "intuitive" or "reasonable" or "right" given the symmetry.

Cognitive moves such as these—in which the mind goes from observing a symmetry or other structural feature to imposing a probability distribution—are traditionally thought to be justified, if at all, by something that philosophers now call the *principle of indifference*. The principle is a rule of right reasoning, an epistemic norm; its treatment thus falls into the domain of probabilistic epistemology, broadly construed. It is with logical or philosophical methods, then, that I will examine the possibility that the force of Maxwell's microequiprobability posit, and so the power and truth-conduciveness of his argument, rests on indifference.

The difficulties in interpreting the principle have prompted many epistemologists to give up on indifference altogether. That, I hope to persuade you, is a bad mistake. There are important forms of thought in both the sciences and in everyday life that turn on reasoning from symmetries and other structural properties to probabilities; if we do not understand such reasoning, we do not understand our own thinking.

That said, the principle of indifference is a chimera—a fantasy, but a fantasy made up from real parts. For me, the more important of these parts is a rule warranting the inference of physical probability distributions from physical structure; this, I will argue, is what gives Maxwell's microequiprobability posit its foundation. But the other part deserves attention too; I will have something good to say about it in passing.

3.2 Classical Probability and Indifference

The historical principle of indifference, then known as the principle of insufficient reason, was born twinned with the classical notion of probability, canonically defined by Pierre Simon Laplace as follows:

> The theory of chance consists in reducing all the events of the same kind to a certain number of cases equally possible, that is to say, to such as we may

be equally undecided about in regard to their existence, and in determining the number of cases favorable to the event whose probability is sought. The ratio of this number to that of all the cases possible is the measure of this probability, which is thus simply a fraction whose numerator is the number of favorable cases and whose denominator is the number of all the cases possible (Laplace 1902, 6–7).

The definition of classical probability and the historical principle of indifference are one and the same, then; consequently, if you apply the principle correctly, you will have certain knowledge of the relevant classical probability distribution. Assuming, for example, that it is "equally possible" that each of the dodecahedral die's faces ends a toss uppermost, you may apply the principle, or what amounts to the same thing, apply the definition of classical probability, to gain the knowledge that the classical probability of obtaining a 5 is 1/12.

The classical definition, and so the indifference principle, raise a number of difficult philosophical questions. First, Laplace's two descriptions of the conditions under which the definition can be applied—to "equally possible" cases, and to cases "such as we may be equally undecided about"—do not seem to be equivalent. The satisfaction of the latter condition is a matter of mere ignorance, while the former appears to hinge on a feature of the world and not of our knowledge state: immediately before the quoted passage, Laplace talks about our "seeing that" several cases are equally possible, and later he writes that, in the case where outcomes are not equipossible, it is necessary to "determine . . . their respective possibilities, whose exact appreciation is one of the most delicate points of the theory of chances" (p. 11). As many writers have observed, classical probability itself inherits this apparent duality in its defining principle (Hacking 1975; Daston 1988).

Second, it is clear that some differences between cases are irrelevant to the application of the principle. That the faces of the dodecahedral die are inscribed with different numbers ought not to affect our treating them as "equally possible" or our being "equally undecided" about them. Why not? What properties are and what properties are not relevant to determining the symmetries and other structural properties that fix the distribution of probability?

Third, Laplace's definition apparently makes it possible to derive knowledge of the classical probability of an event on the grounds of personal ignorance concerning the event. The less you know about a set of events, it appears, the more you know about their probability. This peculiarity seems positively objectionable if classical probabilities are allowed to play a role in prediction or explanation. So Ellis, for example, objects to Herschel's basing a probability distribution over measurement error on our ignorance of the causes of error, but then using that distribution to predict—successfully!—the actual distribution of errors, that is, the frequencies with which errors of different magnitudes occur (section 2.2). *Ex nihilo nihil,* as Ellis (1850, 325) quite reasonably writes. The same objection can be made to certain of Laplace's own uses of the principle.

These problems, though well known, have been eclipsed by a fourth, the late nineteenth-century demonstrations that the indifference principle fails to deliver consistent judgments about the probability distribution over real-valued quantities, a result that has come to be known as Bertrand's paradox (von Kries 1886; Bertrand 1889).[1] A simple version of the paradox—a variation on Keynes's variation on von Kries—is due to van Fraassen (1989, 303–304). Consider a factory that produces only cubes, varying in edge length from 1 centimeter (cm) to 3 cm. What is the probability of the factory's next cube having edge length less than 2 cm? It seems that the indifference principle will advise me to put a uniform distribution over edge length, arriving at a probability of 1/2. But in my state of utter ignorance, I might surely equally well put a uniform distribution over cube volume. A cube with edge length less than 2 cm will have a volume between 1 and 8 cubic cm, while a cube with edge length greater than 2 cm will have a volume between 8 and 27 cubic cm. A uniform distribution over volume therefore prescribes the answer 7/26. What is the probability of a cube with edge length less then 2 cm, then? Is it 1/2 or 7/26? The principle of indifference seems to say both, and so to contradict itself.

To catch the indifference principle in this Bertrandian trap, two assumptions were made: that the principle recommends a probability distribution in any circumstances whatsoever, even given the deepest ignorance, and that the principle cannot recommend two or more distinct distributions.

Some supporters of indifference have rejected the first assumption, holding that the principle can be applied only when a certain unambiguous logical or epistemic structure has been imposed on the problem (Keynes 1921; Marinoff 1994; Jaynes 2003). Jaynes asks, for example, with what probability a straw "tossed at random" and landing on a circle picks out a chord that is longer than the side of an inscribed equilateral triangle, a question first posed by Bertrand himself. His answer turns on the problem's pointed failure to specify the observer's precise relation to the circle. From this lack of specification, Jaynes infers that the probability distribution over the straw's position is translationally invariant in the small, that is, identical for circles that are very close—a conclusion that he remarks is in any case "suggested by intuition." This and other mathematical constraints implied by the problem statement's meaningful silences uniquely determine an answer to Bertrand's question, dissolving the paradox in one case at least. What is more, the answer corresponds to the observed frequencies. *Contra* Ellis, indifference reasoning therefore allows, Jaynes intimates, prediction of the frequencies "by 'pure thought'" (Jaynes 2003, 387).[2]

The second assumption used to generate the Bertrand paradox, that the indifference principle must supply a unique distribution, can also be rejected. Suppose that the function of the principle is to provide, not the truth about a certain objective probability distribution, but a distribution that is a permissible epistemic starting point given a certain level of uncertainty. If a range of such starting points are allowed by the canons of rationality to thinkers in a given state of ignorance—as Bayesians, for example, typically assume—then a principle specifying what those canons have to say to uninformed reasoners will offer up more than one and perhaps very many distributions, from among which the reasoner must freely choose.[3]

3.3 Splitting Indifference

After Laplace, the notion of probability began to bifurcate. The separation is now complete. On the one hand, there is physical probability, the kind of probability that scientific theories ascribe to events or processes in the world, independently of our epistemic state. Physical probabilities predict and explain frequencies and other statistical patterns. They are usually understood to include the probabilities attached to gambling setups such as tossed dice and roulette wheels, the probabilities found in stochastic population genetics—and the probabilities of statistical physics.

On the other hand is epistemic probability. It represents not a state of the world but something about our attitude to or knowledge of the world, or alternatively the degree to which one piece of information inductively supports another, regardless of whether the information is accurate. Epistemic probability is most familiar today in the form of the subjective probabilities found in Bayesianism and other probabilistic epistemologies; as such, it represents an attitude toward, or the relative amount of evidence you have for, a state of affairs' obtaining. The probabilities that epistemologists assign to scientific hypotheses constitute one example; another is constituted by the probabilities you assign to the three doors in section 3.1—they represent something like your current knowledge about, not the observer-independent facts about, what the doors conceal.

If you want to represent the dynamics of jury deliberations, then you will use epistemic probability. If you want to calculate advantageous odds for a gambling game, you will use physical probability. The classical notion was used for both, but the resulting conceptual strain was ruinous. Physical probability and epistemic probability are now widely agreed to be two quite different sorts of thing (Cournot 1843; Carnap 1950; Lewis 1980).[4]

The principle of indifference, which came into the world along with classical probability, should bifurcate too—so I will argue (following Strevens 1998 and North 2010). There are two "principles of indifference," then. Each fits Laplace's schema, in the sense that it takes you from symmetries and other structural features to probability distributions. The principles differ in just about every other way: in the kinds of structure that they take into account, in the kind of probability distribution they deliver on the basis of the structure, and in the sense of "rightness" that they attribute to the distribution in light of the structure.

The first principle of indifference is a physical principle: it leverages your knowledge of the physical world to provide more knowledge of the physical world. The probability distributions it delivers are physical probability distributions; they represent probabilistic facts that are "out there" and that predict and explain statistical patterns regardless of your grasp of what is going on. The physical principle derives its distributions from the physical structure of the systems to whose products the probabilities are ascribed; use of the principle therefore requires previous knowledge, or at least surmise, as to the physical workings of the process in question. Further, the physical principle aims to deliver the true physical probability distribution; its application is based on the presumption that the physical properties that serve as inputs are sufficient, or at least are sufficient often enough in the usual circumstances (whatever they may be), to determine the actual physical probabilities of the corresponding events.

The paradigmatic application of the physical principle is to a gambling device such as the tossed dodecahedral die. The physical symmetries of the die are enough, given certain rather general facts about the workings of the world, to fix the probability distribution over the faces of the die (as I will later show); the validity of the physical principle turns on this connection.

The second principle of indifference is an epistemic principle, not just in the sense that it tells you how to reason (both principles do that), but in the sense that its subject matter is right reasoning. It takes as its premise your epistemic state, with special attention to symmetries in your knowledge, or rather in your ignorance, about a certain set of events. It then delivers an epistemic probability distribution over those events that reflects what you do and do not know about them. The physical indifference principle is all about the world; the epistemic principle is all about you. Insofar as the epistemic probabilities it endorses are "correct," that correctness does not consist in some correspondence to the world, but rather in the aptness of the probabilities given your level of ignorance. Either the probabilities directly represent that state of ignorance, by distributing themselves so as to make no presuppositions about the world where you have no grounds to make presuppositions, or they are epistemic probabilities that you can reasonably adopt, given your level of ignorance. (Perhaps these are two facets of the same notion of what is epistemically appropriate or fitting.)

The paradigmatic application of the epistemic principle is to the case of the three doors. You have no reason to think that the tiger is behind one door rather than another. An even allocation of probability among the doors reflects this state of epistemic balance.

Why think that the case of the doors is governed by a different principle than the case of the die? The probability distribution based on the physical symmetries of the die is a powerful predictive tool, and you know it: merely by applying the physical principle, you can be confident that you are in a position to predict patterns of outcomes in die rolls with great success (if your assumption that the die is well balanced is correct). The probability distribution based on the epistemic symmetries in the game of doors does not have this property. Though you know you have applied the principle correctly, you also know that you should not expect actual frequencies to match the probabilities so obtained (assuming for the sake of the argument that you are highly resistant to tiger bites, so able to play the game many times over).

Once indifference is split down the middle, and two distinct principles with separate domains of application are recognized—what I have been calling the physical principle and the epistemic principle—the objections surveyed above to reasoning from symmetries and other structures to probabilities evaporate.

Taking these difficulties in reverse order: the fourth, Bertrand's paradox, should be handled differently by the two principles. The physical principle will not recommend a probability distribution in cases such as the cube factory because there is not enough information about the physics of cube production to go on; for example, there are no known physical symmetries in the cube-makers.

The epistemic principle may or may not make a recommendation. But if it does, then I suggest that it should not try to emulate its physical counterpart by endorsing a unique probability distribution over edge length, let alone trying to give a distribution that will accurately predict frequencies (as would-be solvers of Bertrand's paradox sometimes seem to be doing). Rather—so my Bayesian instincts tell me—it should suggest a range of possible distributions, proposing that any of them would be a reasonable probabilistic starting point in the circumstances. Among the reasonable candidates will be uniform distributions over both edge length and cube volume. Not included, I imagine, will be a distribution that puts all its probability on an edge length of 2 cm, because such a distribution is unreasonable given our ignorance of the cube factory's workings—or, if you like, it fails to represent our lack of information about such workings.

Defenders of the principle in its epistemic guise have seldom, I should acknowledge, taken this line. Pillars of objective Bayesianism such as Jeffreys (1939) and Jaynes (2003) insist that the principle recommends a single correct prior probability distribution, or at least a single correct prior ordering of hypotheses, rather than permitting a range of reasonable choices. The most recent literature seems also, overwhelmingly, to suppose that the epistemic principle purports to recommend uniquely rational distributions (Norton 2008; Novack 2010; White 2010).[5] Perhaps this is, in part, due to a failure to distinguish the two principles, so demanding from one the uniqueness that is essential only to the mission of the other.

The third objection was that of Ellis and other empiricists: the use of the indifference principle by Herschel, the classical probabilists, and their fellow travelers to gain predictive knowledge of frequencies from ignorance alone violates some grand epistemic conservation law. The empiricists are right. There is a principle that recommends probability distributions on the basis of ignorance alone, and there is a principle that recommends probability distributions with predictive power. They are not the same principle.

What are the symmetries, and more generally the structural properties, on which the indifference principle should base its recommendations? That is the second problem. Again, the solution depends on which of the two principles you have in mind. Much of what follows in this book is an attempt to answer the question for the physical principle. You can at least see, I hope, that principled answers are possible: the numbers on the face of the dodecahedral die are not symmetry-breakers because they make no difference to the physical dynamics of the die.

The first objection to Laplace's characterization of the indifference principle was that it seemed simultaneously to be about the world and your mind, or about facts about equal possibility and facts about your undecidedness. No wonder; it is two principles twisted into one.[6]

In the last century or so, the philosophical and allied literature on the principle of indifference has recognized the dual aspect of the principle in one way and denied it in another. It has recognized the duality by giving the principle a more purely epistemic cast than it finds in Laplace and other classical probabilists. Writers such as Jeffreys (1939) and Jaynes (2003) emphasize that the principle recommends epistemic probability distributions on the basis of the user's epistemic state, which distributions are supposed to be the best representation of the user's information or lack thereof.

Yet some of the same writers have failed to recognize the existence of an entirely distinct physical principle. As a consequence, cases such as the dodecahedral die are treated by both friends and enemies of indifference as paradigmatic applications of the epistemic principle. Jaynes' solution of Bertrand's chord problem, for example, fails to countenance the possibility that our intuitive judgments about the probability distribution's symmetries draw on the physical structure of the chord-determination process, which would make it less mysterious that the solution matches the observed frequencies. Jaynes also cites Maxwell's derivation of the velocity distribution as a classic application of the epistemic principle to obtain predictive probabilities. The attribution of these empirical successes to the epistemic principle clothes it in glory, but ultimately to its great detriment, I think, because it generates impossible expectations about the principle's power to supply unique distributions and because the successes appear so miraculous that the empiricist mind suspects subterfuge or sleight of hand.

Both the physical and the epistemic principles of structure-to-probability inference are real and good. Of the two, I am less sure of the logical status of the epistemic principle, although I do not doubt that it has a psychological grip: in conditions of deep ignorance, some distributions, relatively uniform with respect to the possibilities as they are typically individuated, seem far more reasonable than others. In any case, the epistemic principle will now, for the duration of the book, be put aside.

What I am after is the true form of the physical principle, since it is the physical principle, if anything, that will explain the rhetorical force and the truth-conduciveness of the microequiprobability assumption—the probabilistic posit that grounds Maxwell's alternative, unofficial, underground derivation of the velocity distribution. Looking ahead further, the physical principle also grounds, I will argue, the structure-to-probability inferences that drive discovery in many other scientific domains.

A terminological interstice: "principle of indifference" is an apt name for a rule concerned with the ramifications of ignorance, less so for a rule allowing the inference of physical probabilities from physical structure. Let me therefore give physical indifference a new name: I call the physical principle, which as you will soon see is not a single rule but a suite of interlocking but distinct methods, *equidynamic reasoning* or *equidynamics*. It has not yet earned this name, since I have not yet shown that the inferences in question have anything to do with the causal properties, hence the dynamic implications, of the relevant physical structure—but care will be taken of this matter soon enough.

What do we need to know about equidynamics? We need to know its scope: to what kinds of systems can it be applied? Gases and gambling systems at least, I hope. We need to know its form: what symmetries and what other structural properties does equidynamic reasoning take as its premises? Given these premises, what physical probability distributions does it yield as its conclusions? And we need to understand how equidynamic reasoning is justified—that is, why the close relationship between physical structure and physical probabilities presumed by equidynamics exists.

But first you might ask: would Maxwell and his readers have recognized the force of equidynamic reasoning? If I am right that probabilistic thinkers from Laplace (and before) through Jeffreys and Jaynes have generally conflated the epistemic and physical principles, what reason is there to suppose that Victorian scientists would consider Maxwell's posits to be a solid basis for deriving facts about the actual distribution of molecular velocities?

Could our probabilistic instincts outrun our probabilistic philosophy?

4

THE COGNITIVE WAY

4.1 Balls and Babies: Sampling

Take an eight-month-old infant; sit it down in front of an urn full of balls, mostly white but a few red. Draw five balls from the urn. The infant will expect your five-ball sample to consist mostly or entirely of white balls; they will be surprised if you draw mostly red balls. It seems, then, that based on their observation of the composition of the urn and the sampling procedure alone, and with no additional information about the frequencies with which balls of different colors are drawn, the infants will act as though the probability of a white ball is high and that of a red ball is low, and as though the individual drawings are stochastically independent. Or to put the same claim more compactly, they will act as though they ascribe a *Bernoulli distribution* to ball-drawings, with the probability of white and red based on the proportions of balls of those colors in the urn—just as Laplace recommends.[1]

You will have some questions. How do we know this? Is there any evidence that the infants really do ascribe a Bernoulli distribution to ball-drawings, rather than merely acting that way? And is there any evidence that the distribution is ascribed on the basis of physical properties of the ball-drawing setup?

The experimental paradigm that demonstrates the urn-sampling effect was devised by Xu and Garcia (2008). (Using a slightly different experimental setup, also involving sampling from urns, Denison et al. [2013] have found the same effect in six-month-olds.) The "urn" is a clear plexiglas box filled with ping-pong balls. The drawings are conducted by an experimenter who shakes the box, closes her eyes, and reaches into the top of the box to pull out the balls (in fact drawing from a secret compartment). The infants are tested, after familiarization with the balls and the box, in two ways.

In the first test, the familiar box is put away and another box brought out, covered so that its contents cannot be seen. A sample of five balls is drawn from the box; samples contain either four white balls and one red ball or four red balls and one white ball. The contents of the box are then revealed: either mostly white balls or mostly red balls. Infants are surprised when the sample does not reflect the population, for example, when a sample containing mostly white balls turns out to have been drawn from a box containing mostly red balls. (You would be surprised too: the probability of obtaining four or five white balls from a box containing sixty red balls and twelve white balls is less than 0.004.)

Surprise is assessed by looking time: as has been established by countless studies in developmental psychology in the last few decades, the more surprised an infant is by a stimulus, the longer they look at it. Experimenters can infer what an infant expects, then, by determining what it looks at least long. In Xu and Garcia's experiments, the infants looked longer at a box whose composition did not match the sample than at a matching box.

The second test is like the first, but with one important difference: infants are briefly shown the contents of the box before sampling begins. They exhibit the same looking-time asymmetry; that is, they expect to see a sample that matches the population, thus they are surprised by an unrepresentative sample. In short: given a population, infants expect a representative sample, and given a sample, infants expect a matching population.

What reason is there to think that infants are ascribing independent probabilities to individual events?

In a different kind of experiment, Denison and Xu (2010b) showed infants two jars, each containing a mix of two kinds of lollipops, some pink and some black. The experimenter had previously determined which of the two the infant preferred, pink or black. One jar contained mostly pink lollipops, one mostly black lollipops. A lollipop was sampled from each jar (using the same procedure as above), and the samples were put in separate cups, in such a way that the infants knew which sample had gone into which cup but without observing the sample's color. They were encouraged to crawl toward the cups to retrieve a lollipop. The experimenters observed which cup the infants chose to search first. Infants preferred the cup containing the lollipop drawn from the population in which their preferred color predominated.[2] (I should note that the subjects were older than in Xu and Garcia's study; the protocol requires subjects capable of self-propulsion.) It appears, then, that infants are attaching something probability-like to individual outcomes.

Do the infants assume that samples drawn from a box of balls or a jar of lollipops are stochastically independent? Do they assign equal probabilities, for example, to all samples containing four red balls and one white ball, regardless of the position of the white ball in the line-up? There is some evidence (Xu, personal communication) that they are subject to the gambler's fallacy, expecting later draws to make up for earlier draws so as to ensure a representative sample.

Of course, adults make the same mistake (Croson and Sundali 2005). Despite this fallacy, adults seem nevertheless to assume a "rough independence" that predicts frequencies with approximately the probabilities assigned by a genuine Bernoulli distribution, but with a slight bias toward sequences where later outcomes "save" the representativeness of a sample. (In section 5.1, I will amend this suggestion, proposing that the gambler's fallacy is driven not by a single cognitive principle of "rough independence," but rather by a commitment to genuine independence muddied and unsettled by various cognitive crosscurrents.)

Is there evidence for infants' making an assumption of independence, at least of the rough variety? There is indirect evidence, since you need to assume something like independence—at least tacitly—to get from single-event probabilities to a probability distribution over larger samples that predicts frequencies roughly equal to probabilities. If infants' expectations about frequencies are based on single-outcome probabilities, then, it seems likely that they assume at a minimum rough independence.

There are alternative explanations of Xu and Garcia's results, however. Perhaps infants' expectations about larger samples are computed in parallel with, not on the basis of, their expectations about single outcomes. Or perhaps, when conditions are right, infants follow the rule: expect frequencies approximately equal to proportions of colored balls. For the degenerate case where the sample size is 1, they expect a frequency of 1 for the more populous ball and of 0 for the other, since this is the closest to "approximately equal" that you can get.

I see two ways to empirically distinguish these suggestions. The first is to investigate cases where all probabilities are less than one-half, for example, where there are four colors of lollipop in proportions 2:1:1:1. If infants can be shown to prefer a sample drawn from a jar in which their favored flavor is better represented than the others, that would suggest that they are thinking in terms of probabilities and not merely in terms of representative frequencies. The second experiment would mix legitimate samplings of individual balls with illegitimate samplings in which, for example, the experimenter is clearly not choosing randomly (see below). Infants who think in terms of single-event probabilities should be better able to handle such cases, since they can build expectations about legitimate subsequences from the ground up (though their complexities might anyway be an insurmountable problem).

Let me leave things there. An independence assumption is a surmise rather than a sure thing, but it does make good sense of the data. There is some reason, then, to suppose that babies think about ball-drawing in a Bernoullian way: they ascribe probabilities to the different possible outcomes of a single drawing, and they combine those probabilities at least roughly in accordance with an assumption of independence to arrive at expectations about frequencies, and perhaps other statistical patterns in larger samples.

Are the probabilities that infants ascribe to ball-drawings physical probabilities? The more pressing and more interesting question, I think, is whether the ascriptions are made on the basis of physical information about the ball-drawing setup.

Xu and Garcia's original study showed that the balls must be drawn from the box to elicit expectations about frequency; when balls were drawn from the experimenter's pocket and placed next to the box, infants did not expect the sample to match the population. The expectation must be based on some fact about the sampling procedure, then, that distinguishes it from pocket withdrawal. Are these facts that license probability ascriptions facts about the balls and the box, or facts about the way in which the experimenter removes the balls from the box? Both, it turns out.

Denison and Xu (2010a) demonstrated that (eleven-month-old) infants take into account whether the balls can be physically removed from the box. They showed infants balls, some of which were covered in Velcro strips that prevented their being removed from their containers. Two kinds of stimuli were then compared.

In the first run of experiments, infants were shown a box containing three kinds of balls, green, red, and yellow. Half the balls were green, and the rest were either mostly red or mostly yellow (in a 5:1 ratio). The green balls had Velcro strips attached, and so were presumed to be immovable.

Samples of five balls were taken from the box, containing either four red balls and one yellow ball or four yellow balls and one red ball. Infants reacted as though the green balls were not in play, that is, as though the box contained only red and yellow balls. They were surprised, for example, when a mostly yellow sample was drawn from a box that, green balls aside, was mostly red, but less surprised when the same sample was drawn from a box where yellow balls outnumbered red balls.

The second run of experiments was identical, except without the Velcro: the green balls were therefore able to be sampled. In this case, infants were just as surprised to see a mostly yellow sample drawn from a box where yellow predominated over red as to see the same sample drawn from a box where red predominated. Presumably, they regarded both kinds of samples as extremely unlikely, given that neither contains green balls despite the fact that half the balls in the box are green.

It seems, then, that infants take into account the physics of the box. Expectations are not based merely on the proportions of balls in the box; the balls must be physically equally available for sampling. (Another effect of this sort will be discussed in the next section.)

Let me now turn to the "physics" of the sampler, that is, the physical and mental state of the experimenter who draws the balls from the box. Xu and Denison (2009) have demonstrated that (eleven-month-old) infants take into account whether the drawing of the ball is performed in an appropriately random manner. Using the two-color setup in which either most balls in the box are white or most are red, they first showed infants that the person drawing the sample had a preference for one color over the other, and then compared two scenarios. In the first scenario, the sampler was blindfolded. In the second, she was not blindfolded and she looked into the box when removing the balls. Infants' expectations for the blindfold trials were as normal: they expected the sample to match the population. Without the blindfold, however, they expected the sample to match the sampler's preferences, not the population; for example, if the sampler had a preference for red balls, they expected a sample consisting of red balls even when white balls predominated in the box.

It would therefore appear—rather remarkably—that infants are sensitive to many physical elements that play an important role in randomizing the sampling procedure. All balls must be equally available for selection, and the sampler must not be trying to choose one ball over another. The importance of other elements, most notably the shaking of the box that precedes the selection or any "rummaging around" that the experimenter performs before drawing the ball, is not known at this time.

Although there is further work to be done, then, there is powerful evidence that infants apply some kind of equidynamic reasoning (that is, "physical indifference" reasoning) to urn-sampling procedures: from physical facts about the sampling process and the sampled population, they infer a physical probability distribution over sampling outcomes that has the characteristics, at least roughly, of the Bernoulli distribution, which is of course the correct distribution to ascribe in such cases.

Could Maxwell have based his derivation of the velocity distribution on a physical structure-to-probability principle that all one-year-old humans know how to use? That depends on whether babes' probabilistic wisdom extends from sampled balls to bouncing balls.

4.2 Balls and Babies: Bouncing

Téglás et al. (2007) showed twelve-month-olds a movie of four objects bouncing like gas molecules in a circular container. Three of the objects were identical in shape and color; the fourth was different. There was an exit to the container: a hole at the bottom that was large enough to allow the escape of any of the objects that bounced just so.

After the infants had observed the objects bouncing for a short period of time, the container was occluded, so that they could no longer see what was happening inside. A short time later, one of the objects flew through the exit. In one condition, it was any of the identical three objects; in the other it was the fourth object—the misfit. (The movies used in the experiment are available online at the link given in the references.)

Using looking-time measures, Téglás et al. showed that infants were more surprised when the exiting object was the misfit than when it was one of the three conformists. Apparently they had formed, on the basis of their observation of the setup, the belief that it was considerably more likely that one of the conformists would exit (in the allowed time frame) than that the misfit would exit.

A follow-up experiment demonstrated that the infants' expectations were based at least in part on the physical properties of the setup. Infants were shown a movie much like the first, except that the container was divided in half by a wall. The three conformists were on one side of the

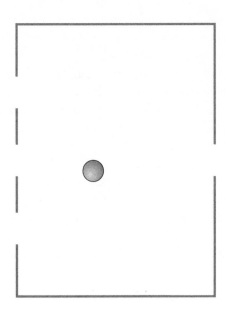

Figure 4.1: A ball bounces around a box with three exits on one side and a single exit on the other; what is the probability that the ball, when it eventually leaves the box, exits the right-hand side rather than the left-hand side?

wall; the misfit and the exit were on the other side. Thus, only the misfit was physically able to exit the container. As before, the inside of the container was occluded and one object exited, either a conformist or a misfit. In contrast to the earlier study, infants looked longer when the exiting object was a conformist.

An interesting variation on the experiment is also reported by Téglás et al. in the same paper. The children in this study were much older (three- and five-year-olds). They saw a single ball bouncing in a container with three apertures large enough for the ball to exit on one side, and a single aperture on the other side (figure 4.1). The container was, as before, occluded; the ball then exited, either on the three-aperture side or on the one-aperture side. Children's expectations were measured not by looking time but by reaction time: they were asked to press a button when they saw the ball exit. They were quicker to detect exits on the three-aperture side, a sign that they thought an exit on the three-aperture side was more likely than an exit on the one-aperture side.

What conclusions can be drawn from these results? It seems plausible that the children in both experiments ascribe something like physical probabilities to the events in question: of a certain kind of object's exiting, and of an object's exiting by way of a certain aperture. There is no evidence for anything more than a qualitative difference in probability. We do not know, for example, that the children ascribe a probability of 1/4 to a misfit's exiting (conditional on any single object's exiting in the allowed time period); nor do we know that they ascribe a precise rather than a vague probability. It seems plausible to suppose, however, that in the first experiment children ascribe equal probabilities to any of the four bouncing objects being the first to exit, and that in the second experiment, they ascribe an equal probability to the ball's leaving via any of the four exits; this would give the obvious probabilities of 3/4 for a conformist's being the first to exit, 1/4 for a misfit exit, 3/4 for an exit from the side with three apertures, and so on.

4.3 Growing Up

And adults? Is there any evidence for their having a facility with equidynamic reasoning to rival that of one-year-olds?

There is much anecdotal evidence, of course. Laplace and every other writer who has had something to say about indifference has supposed that it is natural to attribute equal probabilities to any particular ball's being drawn from an urn (absent physical constraints on the balls, of course, and given that the samplers are not actively selecting particular balls from the urn).

The same goes for the other familiar gambling setups: we effortlessly assign equal probabilities to the two possible outcomes of a coin toss (provided that the coin's mass is evenly distributed), to the several possible outcomes of a die roll (if the die has the form of a regular polyhedron), to the many possible outcomes of a spin of the roulette wheel, and so on. (As I noted in chapter 3, philosophers have, however, been slow to recognize that these are ascriptions of physical probabilities based on the physical structure of the generating setup.)

The psychological literature is less helpful. There is much research on probabilistic reasoning in adult humans, but almost none of it is concerned with ascriptions of physical probability. (In those cases where experiments concern the probabilities assigned to coin tosses, die rolls, and so on, as for example in Johnson-Laird et al. [1999], epistemic and physical probabilities are not distinguished, and so the two kinds of in-difference principle are not separately considered. As a result, the physical bases of the probability ascriptions are not investigated.) I therefore continue in anecdotal mode.

Take Téglás et al.'s bouncing scenarios. You can probably convince yourself, without any formal experimentation, that when there are no physical barriers, adults ascribe roughly equal probabilities to any object's exiting any given aperture (and exactly equal probabilities if the objects are identical). What are their grounds—your grounds—for doing so?

I speculate that you reason as follows. You understand that the movements of bouncing balls and other objects depend sensitively on their exact position. Where a ball will be in a few seconds' time (though not in a microsecond), then, depends not on roughly where it and the other balls are, but on exactly where they are. Further, given that a ball is within a small region, it is equally likely to be anywhere within that region. But then—given some further facts about the dependence of position in the large on position in the small—after a certain amount of time has passed, the ball is equally likely to find itself anywhere within the box. Exit frequencies depend on position. Therefore, any ball is equally likely to exit first (unless its position is constrained in some way), and equally likely to exit through any of the available apertures (if the apertures are physically identical—if they are equally wide, for example).

This story credits you and other adults with considerable probabilistic savoir faire. (On the contrast with humans' famous displays of probabilistic incompetence, as demonstrated by Kahneman and Tversky and others, see section 13.9.) It attributes to you the eponymous part of Maxwell's assumption of the microequiprobability of position (the other part being independence). It also requires you to integrate probabilistic with dynamic thinking, but we know already that even infants do this to some extent, as when they ignore immovable and obstructed balls. Which provokes the thought: perhaps infants reason about bouncing objects in more or less the same, sophisticated way as adults?

I think that this proposal, though not correct in all its details, is close to the truth. Chapter 7 will offer an explanation of infants' and adults' reasoning about bouncing objects according to which equidynamic reasoners make use of all of the elements in the sketch above: sensitivity to initial conditions, facts about what proportion of initial conditions lead to what outcome, and position microequiprobability.

What other applications of equidynamics come naturally to ordinary adults? Without corresponding statistics, and so from nonstatistical facts about causal structure alone, we are able to infer probabilistic facts about interactions between various organisms, about patterns of measurement error in astronomy and elsewhere, about the weather, about other humans, and more—so I will argue in part 3.

4.4 The Birth of Probability

Exaggerating a little, you might say that historians have reached a consensus that the concept of probability—or at least "our concept of probability"—appeared in the West in the seventeenth century. This was classical probability, with its dual nature "both aleatory and epistemological," that is, having to do "both with stable frequencies and with degrees of belief" (Hacking 1975, 10). Midway through the nineteenth century, beginning perhaps with Cournot (1843), the epistemic and physical aspects of classical probability came to be seen as contrary rather than complementary. We ultramoderns are inclined to view epistemic and physical probability as two quite different things that share a mathematical framework.

The literature on infant equidynamics suggests that the historical consensus is deeply wrong. Unless Denison et al.'s six-month-olds, Xu and Garcia's eight-month-olds and Téglás et al.'s twelve-month-olds are remarkably precocious—unless they have drunk down some of the more abstruse conceptual sophistications of the spirit of the age along with their mothers' milk—their equidynamic reasoning cannot be understood as built on a concept that required the intellectual labors of a Leibniz and a Laplace for its introduction into the human mental inventory. It is more likely, surely, that the conception of probability found in infants' equidynamic judgments is a part of the more or less innate toolkit of basic concepts and inferential moves with which the newborn human comes preequipped (Carey 2009).[3]

Why think that this childish idea is a concept of full-blooded physical probability? The notion in question seems likely, if it is to enable the kind of equidynamic reasoning demonstrated by Xu, Téglás, and others, to have the following properties:

1. It is connected to, and perhaps seen as subsisting in, the physical properties of things in the world.
2. It predicts and perhaps explains stable frequencies, and thus of necessity has a corresponding quantitative structure: one probability for every possible frequency.

3. It is applicable, either directly or indirectly, to the single case, as in Denison and Xu's lollipop test.[4]
4. It comes accompanied by a complementary concept of probabilistic independence (also called stochastic independence), a principal lubricant of equidynamic thinking.

That looks very much like the concept of physical probability to me. No doubt it lacks the last degree of philosophical polish and mathematical rigor, but it serves the same inferential function as—and I see no reason to deny that it has the same extension as—the concept of physical probability in the most refined present-day philosophical mind.[5]

In that case, what is the history of "the emergence of probability" all about? It is about a process of self-discovery: the unearthing of the nature of and the building of an explicit, self-conscious model of a category of thought that has existed time out of mind (Franklin 2001, 323). Probability emerged in 1660, then, not *ex nihilo* but from the mental shadows where conscious and unconscious reasoning meet:

> The obstacles impeding the rise of probabilism are not rooted in the facts of everyday life and science. Workable statistics abound, and probability has always been the great guide of life. The difficulties arise out of our attempts to understand and to explain. (Krüger 1987, 83–84)

When did the concept of probability first appear? It appears every day in newborn infant minds.

4.5 The Problem Posed

Maxwell, I have suggested, based his derivation of the velocity distribution for gas molecules in part on the posit that molecular position is microequiprobabilistically distributed, even in gases that are not at equilibrium. The plausibility of that assumption accounts for the powerful, if not overpowering, rhetorical force of Maxwell's paper. The truth of the assumption, meanwhile, accounts in part for the truth of the conclusion; that is, it explains how Maxwell could have derived the correct velocity distribution without any directly relevant statistics.

Why did Maxwell's readers find the microequiprobability postulate—or its consequences, if they did not discern in Maxwell's text the unofficial derivation laid out in chapter 2—to be so reasonable? Because from the age of twelve months or earlier (six months for urns), humans intuitively recognize the force of and actively apply where appropriate equidynamic modes of reasoning that include or imply an assumption of microequiprobability.

Perhaps you find some of this—advanced probabilistic reasoning in six-month-olds, the innateness of equidynamics, and the concomitant innateness of something like the concept of physical probability—hard to swallow. In that case, let me reassure you that, of the psychological theses endorsed in this chapter, the one that matters above all is the universality in space and time of adult equidynamic thinking, which underlies my contention that nineteenth-century physicists, along with the other scientists discussed in later chapters, were sufficiently equidynamically sophisticated to postulate and reason about the microequiprobability of position without assistance. The rest is interesting and important, too, but it is not essential for what is to come.

We are all equidynamicists. The project, then: first, to spell out the form of equidynamic reasoning so as to explain why the ascriptions of physical probabilities that it recommends are typically correct; second, to use these equidynamic principles to understand the persuasiveness and power of Maxwell's derivation. As you will see when I return to Maxwell in section 8.4, equidynamics supplies not only the microequiprobability assumption, but also the other missing pieces in the argument of propositions i through iii, completing the explanation of Maxwell's great armchair discovery.

II

EQUIDYNAMICS

5

STIRRING

Equidynamic reasoning is guided, I will assume, by rules of inference—"principles of physical indifference"—that, though perhaps not directly accessible to introspection, make themselves manifest in the kinds of conclusions about physical probability that we are inclined to draw, from an early age, in the absence of statistical evidence.

The simplest of these rules assigns physical probability distributions to the outcomes of physical processes that have what I will call a *stirring* dynamics. Let me begin with this straightforward case and then treat several more complex cases in turn, building up an account of equidynamics step by step.

5.1 The Wheel of Fortune

Take a disk painted with a large number of equal-sized, alternating red and black sections, arranged like the pieces of a pie, and give it a spin around its axis. Allow it to come to rest, and then see whether a certain fixed pointer is indicating a red or a black section. You thereby obtain an outcome: *red* or *black*. This simple gambling device has been used by a number of writers over the years as a gateway to thinking about the relationship between physical dynamics—in particular, physical symmetries—and physical probability; it is usually called a wheel of fortune (Poincaré 1896; Hopf 1934; Reichenbach 1949; Strevens 2003).

We look at the physics of the wheel of fortune and see immediately that the probability of *red* must be one-half. Such a probability ascription is made with considerable confidence: it would take mountains of statistical evidence to the contrary for us to set it aside.

The inference is not a priori, however, but is rather based on certain physical properties of the wheel, most notably, the symmetry of its paint scheme and of its rotational dynamics. You would abandon your ascription of a one-half probability to *red* if, for example, the red sections were smaller than the black sections, or if you knew that there were small braking pads attached to the red sections that slowed their passage past the pointer relative to the black sections. (You would not cease to think probabilistically, however: you would hold that in the first case, the probability of *red* is less than one-half, and in the second case, it is greater.)

Further, the one-half probability is a physical probability. It quantifies not just something about you—your expectation that a spin of the wheel will yield *red*—but something about the world, namely, the something that explains why the frequency of *red* on a large number of spins of the wheel is typically around 50%.

It seems, then, that the inference of the one-half probability is an application of equidynamic reasoning. Whether infants are disposed to reason in this way about the wheel we do not know (though on the basis of the results reported in chapter 4, we might not be afraid to guess), but certainly adults find the inference easy and natural—in a word, intuitive.

Also intuitive, I think, is a commitment to the independence of the probabilities so ascribed: the adult reasoner believes that the probability of *red* is one-half regardless of what outcomes have previously occurred. Perhaps this claim should be amended to one of "rough independence" in the light of our human weakness for the gambler's fallacy (see section 4.1). Or perhaps we assign independent probabilities, but then in our practical reasoning fail to honor this commitment because of some cognitive infirmity. We might, for example, be good at inferring physical probabilities from physical dynamics, but bad at reasoning from those assignments to more complex facts—for example, to the fact that the sequence BBBB (that is, four consecutive *black* outcomes) is just as probable as BBBR. This suggestion seems rather plausible; I will adopt it as a working assumption.

Suppose, then, that we ascribe to *red* and *black* on the wheel of fortune fully independent physical probabilities, that is, Bernoulli probabilities. But bear in mind that this assumption and the account that rests upon it may have to be qualified in the light of subsequent work on the psychology of human probabilistic inference.

What is the form, and what is the foundation, of the equidynamic reasoning that leads to our ascription of Bernoullian physical probabilities to outcomes of spins on the wheel of fortune? To answer these epistemological questions (building on Strevens [1998]), I will first answer some physical questions: Why is the probability of *red* one-half? And why are spin outcomes independent? That is, what makes the ascription of the physical Bernoulli distribution correct?

As you will see, the answers I give stop short of endorsing any particular metaphysical views about the nature of physical probability; concerning such matters, this book will maintain a strategic ecumenical agnosticism. (Two exceptions: First, I will of course assume that it is permissible to assign physical probabilities to events such as the outcomes of die rolls, the positions of gas molecules, and the trajectories of some organisms, as many writers have argued.[1] Second, in section 12.5 I will briefly abandon my caution, citing metaphysical ideas presented in Strevens [2011] with approval and even, perhaps, a certain self-regarding tenderness.)

5.2 The Probability of Red

Why does the outcome *red* on a spin of a wheel of fortune have a probability of one-half?

Assume that the outcome of a spin is completely determined by the initial speed with which the wheel is spun. (The other parameters affecting the outcome—the wheel's mass and coefficient of friction, for example—are to be considered fixed, and the dynamics of the wheel deterministic; for environmental noise, see section 12.4.) Then it makes sense to talk of the *red*-producing spin speeds, that is, the speeds that, when imparted to the wheel by the croupier, will result in the outcome *red*.

Assume next that there is a physical probability distribution over initial spin speeds; thus, that there is a matter of fact about the physical probability of a given spin's speed falling within any given range. The probability of *red* will then, of necessity, equal the probability of an initial spin speed's falling into the *red*-producing set (assuming that physical probability assignments are equal for events that with nomological necessity co-occur).[2]

What if there is no physical probability distribution over initial spin speed? There may still be a physical probability distribution, or something with just as much predictive and explanatory power, over the outcomes of wheel of fortune spins, I will suggest in section 12.5. It would be distracting to explain my proposal at this early stage, however; let me

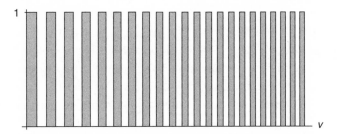

Figure 5.1: The evolution function for the outcome *red* on a wheel of fortune maps initial spin speed v to 1 for values of v that result in a *red* outcome and 0 for values of v that result in a *black* outcome

simply assume that a physical probability distribution over initial conditions exists, leaving it to you to generalize later.

The probability that a wheel of fortune spin yields *red* is equal, you have seen, to the probability that the initial speed of a spin on the wheel falls into the set of *red*-producing speeds. Why should this latter probability equal one-half?

The answer lies in a rather abstract property of the wheel's dynamics. Consider a function that maps all *red*-producing values of initial spin speed to 1 and all *black*-producing values to 0. This is what I call the wheel's *evolution function* with respect to the outcome *red*. (It can also be called the "indicator function" for the property of being a *red*-producing value.)

The function for a typical wheel of fortune is pictured in figure 5.1. The area under the graph is shaded gray; thus, the *red*-producing values of the initial spin speed v are those spanned by the gray bars.

The evolution function has a property that I call *microconstancy:* the range of the spin-speed variable v can be divided into very small intervals, in each of which the proportion of *red*-producing values of v is the same, in this case, one-half. You can think of microconstancy as subsisting in two facts about the evolution function: first, the function oscillates back and forth between zero and one very quickly, which is to say that it would only ever take a small change in initial spin speed to reverse the outcome of a spin on the wheel; and second, the oscillation has, as it were, a regular rhythm. (A formal definition of microconstancy is given in Strevens (2003, section 2.C). Note that microconstancy is relative to an outcome, a way of quantifying initial conditions, and a standard of smallness; section 5.6 elaborates. Let me also remind you that this book contains a glossary of technical terms.)

Because of the microconstancy of the wheel's evolution function, a wide range of initial spin-speed distributions will induce more or less the same probability for *red,* namely, one-half. You can get some intuitive sense of this from figure 5.2. The probability of *red* is equal to the probability of a *red*-producing initial spin speed, which is equal to the proportion of the area under the spin-speed probability density that is shaded gray in the figure. (As in figure 5.1, the gray areas represent *red*-producing initial speeds.) Since the evolution function is microconstant, any reasonably "smooth" and "wide" probability density will have about half its area shaded gray. It follows that, whatever the initial spin-speed distribution, provided it is smooth and fairly wide, the physical probability induced for *red* will be about one-half.

It is for this reason that the characteristics of different croupiers— some spin faster, some slower, some with a higher variance, some with a lower variance—make no difference to the probability of *red,* provided that they spin fast enough (as the casino and bettors will insist they do).

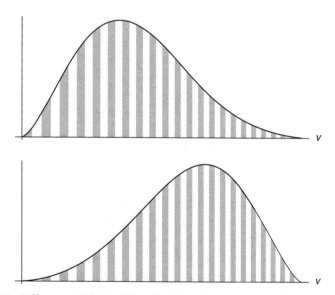

Figure 5.2: Different initial speed distributions induce the same physical probability for *red* of one-half. The probability of *red* is equal to the proportion of the area under the probability density that is shaded gray.

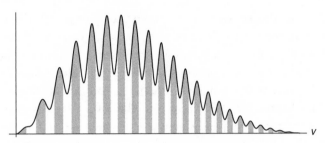

Figure 5.3: A probability density that is not smooth; a probability for *red* that is not one-half

Whoever spins, the probability is one-half. Or roughly one-half: all probabilities in what follows are approximate (though I may not always say so explicitly), but good enough for equidynamic purposes.

It is also why the probability of *red* is the same on different wheels of fortune made of different materials and operated in various conditions. Though the conditions, materials, and methods of construction make a difference to the evolution function, or in other words make a difference to which initial spin speeds map to which outcomes, they do not change the fact that it has the microconstant aspect shown schematically in figure 5.1, inducing a one-half probability for *red* for any smooth, wide initial condition density.

What makes a probability density "smooth"? A sufficient condition for the density to induce a one-half probability is that it be approximately uniform over any small interval (where the standard for "smallness" is the same standard with respect to which microconstancy is defined). That is, the density must not increase or decrease too much as it traverses any neighboring pair of gray and white stripes in figure 5.2. By way of comparison, a nonsmooth density is shown in figure 5.3. As you can see, this unquiet density does not induce a probability for *red* anywhere near one-half—the probability is considerably greater.

In the past I have called this smoothness property *macroperiodicity*. But you may have noticed that it is in essence identical to a property that Maxwell attributed to molecular position, namely, the property of microequiprobability: both properties require that a probability distribution is more or less uniform across any small region of parameter space. (Microequiprobability as characterized in chapter 3 also implies independence; more on this shortly.) From this point on, I use the term *microequiprobability* for the smoothness in question.

Note that because microequiprobability is sufficient for a *red* probability of one-half, the presumption above that the initial condition density is not only smooth but wide—that it spans many alternating gray and white patches—is, strictly speaking, unnecessary. Nevertheless, in real life, we will suspect that a narrow density is not microequiprobabilistic. Thus if a croupier's distribution of initial conditions strikes us as narrow, as it will if the croupier only ever spins the wheel very slowly, or if a robot croupier spins the wheel with almost exactly the same speed every time, we will not expect a probability for *red* of one-half (or perhaps any stable probability for *red* at all). With these remarks I put aside the issue of width as a heuristic for the remainder of the discussion.

The observations I have made about the wheel of fortune are one instance of a general proposition. A definition: if a physical process is microconstant with respect to an outcome—if its space of initial conditions can be divided into many small contiguous regions, in each of which the same proportion of initial conditions produces the outcome—then say that the proportion in question is the process's *strike ratio* for that outcome. (Thus, the wheel of fortune's strike ratio for *red* is one-half, as is its strike ratio for *black*.) Now the proposition itself: if the physical probability distribution over a process's initial conditions is microequiprobabilistic (that is, macroperiodic), and if the process is microconstant with respect to an outcome *e*, then the initial condition distribution will induce a physical probability for *e* approximately equal to the strike ratio for *e*. (A more rigorous statement of this result, with extensions, is proved in Strevens [2003, section 2.C].)[3]

To sum up: for the physical probability of *red* on the wheel of fortune to be one-half, it is sufficient that the physical probability distribution over spin speed be microequiprobabilistic and that the dynamics of the wheel be microconstant with a strike ratio for *red* of one-half. If you come to know that the density and dynamics have these properties, then, you can correctly infer the one-half probability of *red*. It is upon this opportunity that the principles of equidynamics seize.

5.3 Independence

Any two outcomes of spins on the wheel of fortune are stochastically independent. What is the basis for this fact? Let me narrow down the question for expository purposes: what is the basis for the fact that the outcomes of any two consecutive spins are independent?

A sufficient condition for independence is that the initial conditions of the spins are independent, that is, that learning the speed of one spin gives you no information about the speed of the next. But for the wheel of fortune, this condition surely sometimes fails to hold. Suppose, for example, that a certain croupier grows enthusiastic and lethargic in turns, so that the mean spin speed that she imparts to the wheel is greater at some times than at others. Then there is a correlation between successive spin speeds: the probability of a high spin speed is greater following another high-speed spin than following a low-speed spin.

As you might guess, however, this makes no difference to the probability of *red*, and so does not interfere with the independence of the outcomes, provided that the probability distribution over spin speed conditional on any particular value for the immediately preceding spin is microequiprobabilistic, or equivalently, provided that the joint density over successive pairs of spin speeds is microequiprobabilistic, that is, approximately uniform over any small region of the two-dimensional space of pairs of successive spin speeds.

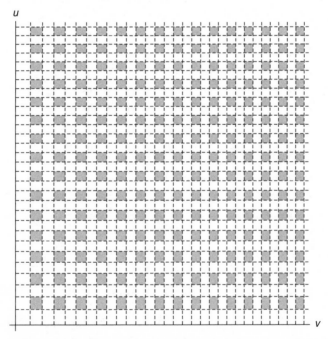

Figure 5.4: The evolution function for the outcome of obtaining *red* on two consecutive spins of the wheel of fortune. The initial speeds of the spins are *u* and *v*; gray areas represent pairs of spin speeds that yield *red* twice.

That the microequiprobability of the joint density is sufficient for independence can easily be discerned by constructing an evolution function for the composite process consisting of two consecutive spins of the wheel and the composite outcome of obtaining *red* on both spins. The evolution function has two initial conditions: the initial speeds of the two spins. It is pictured in figure 5.4; as you can see, it is microconstant with a strike ratio of 1/4. Any microequiprobabilistic density over pairs of successive spin speeds will therefore induce a probability for the double *red* outcome that is also equal (approximately) to 1/4, as independence requires.

The result is easily generalized: to other pairs of outcomes (*BB, BR, RB*), to other numbers of trials (say, three consecutive spins on the wheel), and to other criteria for selecting sets of trials (pairs of outcomes separated by three intervening trials, trials on your birthday, and so on—provided, of course, that the selection criteria do not take into account the initial conditions or outcomes of the trials).[4]

How is it possible that the initial conditions of successive spins are correlated, but that the outcomes entirely determined by those initial conditions are not? The answer is that what are correlated are the approximate magnitudes of the initial conditions, and microconstancy renders these approximate magnitudes probabilistically irrelevant to the outcomes.

5.4 Inferring Microconstancy

In order to infer the existence of a physical Bernoulli distribution (including independence) over the outcomes *red* and *black* on a wheel of fortune, it is sufficient to know that the wheel's dynamics are microconstant with strike ratio one-half and that the joint probability distribution over the initial conditions of any set of trials is microequiprobabilistic.

This creates an epistemic affordance, an opportunity to infer the physical probability of an outcome without examining the statistical patterns with which the outcome occurs. The principles of equidynamics that drive such inferences in both young children and adults take advantage, I propose, of this opportunity. How so? How do we obtain the necessary knowledge of microconstancy and microequiprobability? Knowledge of microconstancy is this section's topic, knowledge of microequiprobability the next's.

It is easy to recognize microconstancy in an evolution function; however, physical systems do not present their dynamics to us in graphical form. What knowledge we have of microconstancy must be based on observations of actual physical configuration and behavior.

The characteristic properties of a microconstant dynamics are, first, a sensitivity of the outcome in question to initial conditions, and second, a certain pattern or symmetry to this dependence.[5] To see that a system has a microconstant dynamics with respect to a certain outcome, then, is to see that the physical properties of the system give rise to such a sensitivity and symmetry.

How is it done? In this chapter, I will focus on the recognition of microconstancy in a particular class of mechanisms, the eponymous "stirring" mechanisms. A stirring mechanism is characterized by a physical variable with a bounded range of possible values that I will call the stirring parameter. Upon activation, the mechanism's stirring parameter begins to cycle smoothly through all its values and continues to do so, over and over, until it is stopped. The value of the stirring parameter at that point determines the outcome—*red* or *black,* heads or tails, or whatever. In this context, "smooth" means that the speed of the cycling changes, if at all, quite slowly, so that the time taken for successive cycles is approximately the same.

The wheel of fortune is a stirring mechanism whose stirring parameter is the point on the wheel indicated by the pointer.[6] As the wheel turns, the parameter cycles through every possible value, that is, through every point on the circumference of the wheel. Further, it does so smoothly—the wheel's angular velocity changes only gradually as it slows to a stop. The final value of the stirring parameter—the point on the circumference indicated by the pointer when the wheel is stopped or comes to rest—then, of course, determines the outcome, by way of the wheel's paint scheme.

A stirring mechanism is microconstant if it cycles through the stirring parameter sufficiently many times. The smooth cycling ensures multiple repetitions of a constant strike ratio, that ratio being the proportion of values of the stirring parameter that determine the outcome in question.[7] The "sufficiently many times" ensures that the repetitions are on a sufficiently small scale that the outcome is highly sensitive to the initial conditions.[8]

To recognize that a stirring device has a microconstant dynamics relative to a certain outcome, then, you must recognize that:

1. the device cycles many times through all states of a variable physical property, the stirring parameter,

2. the final value of the stirring parameter determines whether or not the outcome occurs, and
3. the cycling is smooth; that is, the rate of change in the cycling speed changes, if at all, only slowly.

You may then infer that the outcome has a strike ratio, and so a physical probability, equal to the proportion of the values of the stirring parameter yielding that outcome. If one-third of the values of the stirring parameter yield *red* and two-thirds *black*, for example, then you may infer that the probability of *red* is one-third. You may also infer that there is a uniform physical probability distribution over the stirring parameter itself.

To see that a device has properties (1) and (2), you must examine its overall physical configuration. To see that it has property (3), you must appreciate a certain symmetry in its operation. In the case of the wheel of fortune, that symmetry is the rotational symmetry of the wheel's dynamics—that is, the fact that the dynamics is the same whatever the orientation of the wheel. Begin a spin with the wheel in one position, for example, and the wheel will execute just as many rotations, at exactly the same speed and with exactly the same deceleration, as if you had begun with the wheel in any other position. This rotational symmetry can be inferred in turn from the physical symmetry of the wheel itself.

Another stirring mechanism is a simple coin toss, in which a coin is sent spinning into the air and then caught on the fly. (A toss in which the coin is allowed to bounce is slightly more complex; see section 6.1.)[9] The tossed coin's stirring parameter is the angle between the coin and the ground; the coin cycles smoothly through values of this parameter between 0 and 360 degrees, and does so many times in the course of a toss. Consequently, the coin has a microconstant evolution function, as shown in figure 5.5.

As with the wheel of fortune, the tossed coin's microconstancy can be inferred from the overall configuration of the setup and the physical symmetry of the coin itself, which is responsible for its smooth rotation and hence for its cycling smoothly through the stirring parameter.

Let me take stock. I have shown that the microconstancy of a stirring process can be inferred on the basis of relatively few, large-scale, and therefore typically accessible properties of the process. Naive or folk physics is sufficient to see that the spinning wheel of fortune and the tossed coin are microconstant with respect to *red* and heads, and that the strike ratios in both cases are equal to one-half.

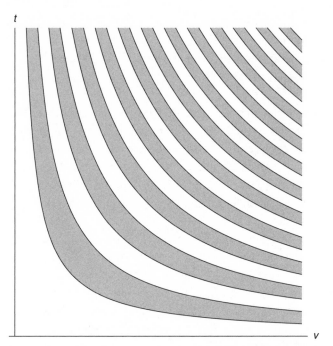

t

v

Figure 5.5: Evolution function for a tossed coin with two initial conditions, the coin's initial spin speed *v* and the time *t* for which it is allowed to spin. Pairs of values of *v* and *t* producing an outcome of heads are shaded gray (Keller 1986).

Do humans actually reason about stirring mechanisms in this way? Thinking about tossed coins and wheels of fortune, do we identify a stirring parameter and use it to infer the microconstancy of the evolution function for various outcomes, to calculate the corresponding strike ratios, and so to assign probabilities? That our equidynamic reasoning has this form is a plausible hypothesis.

There is another hypothesis, however, that might explain how we non-statistically infer probabilities for *red*, for heads, and so on. On this view, our reasoning about microconstancy is more heuristic than demonstrative: we infer the wheel of fortune's microconstancy with respect to *red* and *black* from the fact that these outcomes depend sensitively on the relevant initial conditions and the fact that the physical element determining the outcome—the wheel itself—is symmetrical both inside and outside, that is, in its constitution and paint scheme. Such an inference does not identify a stirring parameter and, more generally, it does not explicitly take into account the dynamics of the wheel (except insofar as necessary to determine the sensitivity of outcomes to initial conditions).

Could such an inference go wrong? It would fail if applied to, say, a wheel that is not merely given an initial spin and left to come to rest, but is driven by a device that controls the speed of the wheel in subtle ways, so as to establish a sensitivity to initial conditions but not the kind of regularity that is needed for microconstancy. In an environment without such baroque constructions, however, the inference will be quite reliable. The heuristic therefore offers an attractive trade-off between tractability and accuracy.

There are at least two inference rules that handle stirring mechanisms well, then: the more deeply dynamic rule that identifies stirrers as such, and the sensitivity-symmetry heuristic just described. Which do we use?

An appealing feature of the sensitivity-symmetry heuristic is its provision of a unified treatment of our thinking about stirring processes and of our thinking about what I will call in chapter 6 shaking processes, such as die rolls, drawings from urns, and bouncing balls. But in that chapter, I will also demonstrate a degree of dynamic sophistication in the conclusions we draw about dice, balls, and urns that far outruns in complexity and depth what the sensitivity-symmetry rule can provide, eliminating that rule as an explanation of our thinking about shaking. This suggests to me that our thinking about stirring processes is no less sophisticated, and thus that it is based on the deeper rule that identifies a stirring parameter.

Some terminology to cap things off: I will call the equidynamic rule that we use to infer probabilities for the outcomes of stirring processes, whatever its form, the *stirring rule*.

5.5 Inferring Microequiprobability

Microconstancy is not enough. To infer a physical probability distribution over the outcomes of a microconstant process according to the recipe given by the stirring rule, it is necessary also to infer or assume that there is a physical probability distribution over the process's initial conditions, and that this distribution has the property of microequiprobability.

More exactly, as explained in section 5.3, it is necessary to assume, in order to secure the independence of outcomes characteristic of a Bernoulli process, that there exists a joint physical probability distribution over the initial conditions for any collection of processes—any sequence of coin tosses, for example—and that this joint distribution is microequiprobabilistic. In what follows, this is what I mean when I talk about the microequiprobability of the initial condition distribution.

The question of the existence of the initial condition distribution was earlier deferred to section 12.5. My question here is: assuming that there is a distribution, on what grounds might we infer or suppose that it is microequiprobabilistic?

A natural thought is to apply equidynamic reasoning to the process generating the initial conditions. But whereas we have a reasonable grasp of the dynamics of the wheel of fortune, we know very little about the mechanics of the process that spins the wheel. It is quite familiar, of course—we ourselves are the spinners—but its inner physical properties are obscure to us or, certainly, to most of us. It therefore seems unlikely that the attribution of the probability of one-half to *red* is based on my, or your, or some six-month-old's understanding of the process by which the human hand imparts a variable speed to the wheel of fortune. Perhaps a sophisticated physiologist—the Maxwell of motor control—could reason in this way (see section 12.2), but we ordinary humans must be following some other line of reasoning.

What are we thinking?

A first possibility is that, having spent years in close proximity to our hands, we are familiar with their statistical output. We see that the distribution of spin speeds, or of some broader class of actions that includes spin speeds, is of the sort characteristically produced by a microequiprobabilistic probability distribution, and we consequently infer that our bodies are built so as to produce spin speeds microequiprobabilistically.

Unlikely, I think. It is far from clear that I am able to measure the magnitudes of my hand movements and other bodily motions precisely enough to see that their distribution is microequiprobabilistic. Even if I do have so fine an eye, the hypothesis requires extraordinary mnemonic prowess and attention to detail.

A second and related possibility is that I infer backward from the observed statistics of coin tosses and so on to the microequiprobability of their initial conditions. I know from coin tosses' conforming to Bernoulli statistics, for example, that they likely fall under a physical Bernoulli distribution. The best explanation of this distribution is the hypothesis that the dynamics of tosses are microconstant (which I can confirm independently) and that their initial conditions are distributed microequiprobabilistically. So I infer that the initial conditions are so distributed, and enlarge my inference to include other similar bodily actions, such as spins of the wheel of fortune, for whose distribution I may lack direct evidence. Other processes perhaps provide even better evidence for the microequiprobabilistic tendencies of the human body—for example, if I am a keen darts player, I may be able to see in my patterns of near misses a distribution that is nearly uniform over small regions.

Some people are surely in a position to reason in precisely this way. But what if you do not play darts, or attend very much to the statistics of coin tosses? What if you are six months old? If this hypothesis as to our reasons for supposing the microequiprobability of spin speed were correct, there would surely be far more variability in humans' willingness to reason equidynamically than we in practice observe.

A third possibility is that, rather than inferring microequiprobability in particular cases, I have a blanket policy of defeasibly assuming things to be microequiprobabilistically distributed unless there is some positive reason to refrain (as when, for example, the statistics do not bear out the microequiprobability posit, or when I have a device that converts a microequiprobabilistic distribution over its inputs into a nonmicroequiprobabilistic distribution over its outputs). According to this hypothesis, I simply assume that wheel of fortune spin speeds are microequiprobabilistic; no further thought is required. Such a strategy may work rather well in practice, if microequiprobabilistic distributions are the rule.

The fourth and final hypothesis is a variant on the third: we might assume microequiprobability not for everything but for some circumscribed set of processes. We might, for example, suppose without argument that human bodily motions are distributed microequiprobabilistically. Perhaps such assumptions are innate; that would help to explain infants' facility with equidynamics. Various tests are possible: would infants be more reluctant to engage in such reasoning if a mechanical, rather than a flesh-and-blood, arm were plucking colored balls from a box?

Such are the options. There is no need to choose among them now. In chapter 12 I will return to the question, endorsing versions of both the third and fourth hypotheses and suggesting that they are only tacit in the stirring rule, so that a user of the rule can assign probabilities to the outcomes of a stirring process without having to think explicitly about the distribution of initial conditions, hence based on the dynamics of stirring alone. Let me finish by emphasizing that on any of these approaches, inferences of physical probabilities for the outcomes of stirring processes will be defeasible, because while there can be justified confidence, there can be no certainty about initial conditions' microequiprobability.

5.6 Relativity

Microconstancy is relative to an outcome, a way of quantifying the initial conditions, and a standard for what is "micro," as is the concomitant notion of a strike ratio. Microequiprobability is relative to a way of quantifying the initial conditions and a standard for "micro." Uniformity (of a probability distribution) is relative to a way of quantifying the distributed property.

Throughout this book, I economize by omitting to recognize explicitly some or all of these relativizations; I leave it to you to say the words mentally that are missing on the page. This section exists to comment briefly on the epistemic and metaphysical implications of relativity.

Let me begin by explaining what it means for a property such as microconstancy to be relative to a way of quantifying initial conditions. Consider molecular speed (the magnitude of molecular velocity), a scalar physical quantity. There are infinitely many ways to represent such a quantity using the real numbers. Each such way corresponds to a function from speeds—the actual, worldly physical properties—to the real numbers. Measuring speed in centimeters per second corresponds to a different function than measuring speed in meters per second, for example: the speed mapped by the first function to 100 is mapped by the second function to 1. In the probability literature, such functions are called *random variables*.[10] (The term makes a certain amount of sense, yet it invites confusion, first because a random variable is not a variable but a function, and second because there is nothing intrinsically random about a random variable.)

The difference between speed measured in meters per second and speed measured in centimeters per second is philosophically uninteresting. But random variables need not be linear functions of each other. You can, for example, measure speed in meters per second squared, by representing a speed of v ms^{-1} using the number v^2.[11]

The existence of different quantification schemes matters because properties such as uniformity are relative to a scheme. If the probability distribution over speed measured in meters per second is uniform, then the distribution over speed measured in meters per second squared is not uniform, and vice versa. (For an illustration of the relativity of microequiprobability, see section 12.1.) Further, if a quantity's probability distribution satisfies a certain broad criterion—absolute continuity, that is, representability by a probability density—then there is guaranteed to exist a quantification scheme relative to which it is distributed uniformly (over those regions where its probability is nonzero). Similarly, for any evolution function satisfying a related criterion, there is guaranteed to exist a quantification scheme relative to which the function is microconstant for a given outcome with any strike ratio you like.

This raises the question: which quantifications should we care about? Relative to which quantifications does microconstancy have significant implications?

To this question I give a standard answer: you should use what I call *standard variables*—roughly, variables that measure physical quantities in proportion to the SI units, such as meters per second. The epistemic and metaphysical significance of the standard variables is not that they have been approved by an international committee, but rather that they tend, under a wide range of circumstances, to adopt a microequiprobabilistic distribution (section 12.3).

Microconstancy with respect to a quantification of the initial conditions predicts and explains frequencies only if the corresponding initial condition distribution is microequiprobabilistic with respect to the same quantification. Thus, microconstancy with respect to a standard variable is interesting, because you can assume that such a variable is microequiprobabilistically distributed, while microconstancy with respect to a highly nonstandard variable is not, since the microequiprobability of such variables' distributions cannot safely be assumed. It follows that knowledge that a process is stirring, hence microconstant with a certain strike ratio, relative to a standard variable affords an epistemic opportunity that does not exist for stirring relative to a highly nonstandard variable: in the one case microequiprobability can be assumed and a physical probability equal to the strike ratio inferred; in the other case, not.

In these observations lie the essential elements of a reply to Bertrandian worries about stirring and the other rules of equidynamics discussed in this book. Could there be a Bertrand's paradox for equidynamics? Could the rules of equidynamics prescribe two or more inconsistent conclusions about the physical probability of some outcome? No such problems will arise, provided that the rules are applied only with respect to standard variables. I propose consequently that equidynamic rules are limited to standard variables; the notion of a standard variable, then, is an important part of the equidynamic toolkit.

What about the other relativities? The relativity of microconstancy and strike ratio to an outcome needs no comment. The relativity of "micro" is a more important topic. Let me say for now that it is an issue that a sophisticated equidynamic reasoner must to some extent manage by hand, always taking care that there is no mismatch of standards, so that, for example, the "micro" with respect to which a stirring process is judged to be microconstant is the same (or smaller) micro as that with respect to which the process's distribution of initial conditions is assumed to be microequiprobabilistic. The source of some judgments about the standard for "micro" in microequiprobability is discussed in chapter 12; a few further remarks are made in section 13.2.

6

SHAKING

Most stochastic processes are not stirrers. Most gambling devices, even, are not stirrers. The dynamic property of microconstancy is, nevertheless, at the heart of the physical probabilities attached to these other devices (quantum stochasticity aside), and the same patterns of thought that take advantage of microconstancy to infer physical probability distributions over the outcomes of stirring processes may be applied, though with some considerable amendments, to what I will call *shaking* processes.

6.1 Bouncing Coins

Toss a coin as before, but now let it land on the floor, bouncing. The coin's dynamics may be divided into two parts: a stirring phase, and a short bouncing phase. What is the probability of a bouncing coin's landing heads? One-half, of course. Evidently the bouncing phase does not interfere with the probability induced by the stirring phase. Why not?

You might try to analyze the physics of the bouncing phase (Zhang 1990), but it is not necessary. A symmetry argument, formulated succinctly by Engel (1992, 47), shows that a bouncing coin's evolution function is microconstant with a strike ratio for heads of one-half.

The strategy of the argument is as follows. Consider the evolution function for a coin that is tossed, then snatched out of the air an instant before it hits the floor and begins to bounce. The function will have the form shown in figure 5.5. Take any region of initial conditions corresponding, in the figure, to a pair of neighboring gray and white stripes. The aim is to show that allowing the coin to bounce will not alter the fact that about half the region is gray and half white, or equivalently, that about half the initial conditions in the region lead to heads and half to tails. Since the region is microsized, and the argument can be applied to every such region, it will then follow that the space of initial conditions for the bouncing coin can be divided into microsized sets in each of which about half the initial conditions lead to heads, and thus, that the evolution function for the bouncing coin is, as desired, microconstant with a strike ratio for heads of one-half.

How, then, to show that, though bouncing may rearrange the gray and white within any neighboring pair of stripes, it does not significantly change the ratio of gray to white?

For any neighboring gray-white pair in the nonbouncing evolution function, consider a narrow strip of points representing all tosses in which the coin has approximately the same angular velocity at the moment that it is caught (and hence, in the bouncing case, at the moment that the coin lands), as shown in figure 6.1. The tosses represented by this strip correspond to a single, full revolution of the coin. At the bottom of the strip are those tosses where heads has just become uppermost; in the middle, where the gray turns to white, are those tosses that have gone on

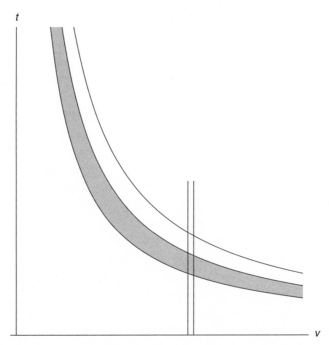

Figure 6.1: The area between the two vertical lines picks out landing conditions in all of which the coin has approximately the same angular velocity v

for slightly longer, so that tails has just become uppermost; at the top are those tosses that have gone on slightly longer still, so that heads is once again about to become uppermost.

Now suppose that the coin is allowed to bounce. The initial conditions in the strip represent tosses in which the coin lands with about the same angular velocity but with every possible orientation, 0 through 360 degrees. Assume that the strip is short enough (measured from top to bottom) that the tosses land with approximately the same translational velocity (for further discussion, see note 1). Then the only physically important difference between the tosses in the strip is the orientation of the coin at the moment that it lands.

Consider the lower half of the strip, corresponding to landings in which heads is facing upward (or at least, closer to facing upward than tails). If there were no bouncing phase—if the coin had been snatched out of the air at this point rather than allowed to bounce—then the result of the toss would have been heads. But the coin bounces. For some landing orientations the bounce will reverse the outcome, producing tails, while for others it will preserve the outcome, producing heads. Compare this with the set of landing conditions in the strip in which the coin starts out with tails uppermost. Because of the physical symmetry of the coin, the dynamics will be the same as in the heads case; thus, the same orientations will reverse the outcome—in this case, from tails to heads— and the same orientations will preserve it. Consequently, the proportion of heads-uppermost landings that are reversed rather than preserved will exactly equal the proportion of tails-uppermost landings that are reversed rather than preserved. The bouncing phase will therefore not change the proportion of heads-producing initial conditions in the strip.

Repeat this argument for every strip in every neighboring gray-white pair, and you have what you want: microconstancy of the evolution function for a bouncing coin with a strike ratio of one-half.[1]

In short, you can reason about the bouncing coin as follows: the stirring phase of a coin toss has the kind of microconstant dynamics that produces heads or tails with equal probability; the bouncing phase preserves this aspect of the dynamics, and so produces heads and tails with the same probabilities. (You need not assume, but you ought to leave open the possibility, that the bouncing phase makes a positive contribution to the bouncing coin's microconstancy; it is possible, for example, that the bouncing of some coins is itself a stirring process.) I call this sort of reasoning *proportional dynamic* reasoning: it involves thinking quantitatively, but not explicitly probabilistically, about what proportion of initial conditions will result in what outcome.

You *can* reason as follows, I said, but do you? When inferring the probability of heads from the physical symmetry of the bouncing coin, do you run through the proportional dynamic argument above to convince yourself that bouncing preserves microconstancy? Or do you rely on a heuristic of the form described in section 5.4, a heuristic that advises: where there are physical symmetries and sensitive dependence on initial conditions, infer proportional physical probabilities? To answer the question, I need to uncover more of the substructure of equidynamics.

6.2 Tumbling Dice

An adult human reasoner ascribes equal probabilities to the outcomes of a die roll just as easily as they ascribe equal probabilities to the outcomes of a coin toss. The probability-inducing dynamics of the die roll are far more difficult to analyze, however, than the stirring dynamics of a coin toss. Human reasoners therefore appear to be taking a short cut, or so I will argue. As before, I first discuss dynamics without regard for epistemology or psychology, and only then turn to the question how the mind gets its grip on the probabilistic world.

No single analysis will capture the probabilistically relevant dynamics of every die roll, I think: there are variations in how the dice are handled and thrown, and in the surfaces on which they tumble, that contribute in qualitatively different ways to the physical probabilities in different circumstances. But an analysis of a common kind of roll will provide quite enough material to go on.

Take a well-balanced cubical die, then; cup it in both hands and give it a good shake; toss the die in a sweeping lateral motion onto a rug or similar surface; let it roll to a stop. What is the probability of obtaining a 5? It is of course the same as obtaining any of the other five possible outcomes: one-sixth.

The reason, I will suggest, is that the evolution function for rolling a 5 is microconstant with a strike ratio of one-sixth. As in a stirring process, then, the probability in question may be inferred from the properties of the physical dynamics that make for microconstancy and an assumption of initial condition microequiprobability, without consulting statistics. But the die roll is not stirring or, at least, not in the right way.

The roll has three distinct parts: the initial shaking of the die, its flight through the air, and its bouncing upon landing. Begin with the middle phase. Since the die is spinning freely, experiencing no forces other than that of gravity, it will move much like the coin, rotating rapidly around an axis determined by its state when it is released from the hand.[2] The result is an action much like the coin's stirring, with one important qualification: the die will not, and cannot, rotate in a way that gives equal time to each face. In the worst case, which is not improbable given the mechanics of throwing, the die will rotate through only four faces, in the sense that there are two faces that are at no time in its flight uppermost— namely, the two faces pierced by the axis of rotation.

What is the effect of the bouncing phase? It depends on the mechanics of the bouncing. In the kind of roll described above, it is quite possible for the die, upon landing, to continue to rotate around its in-flight axis, rolling along the landing surface rather than bouncing "randomly." In this case, there is little or no chance of the two faces on the axis of rotation being uppermost when the die stops. Were the flying and landing phases the only determinants of the outcome of the roll, then, the six outcomes would not be physically equiprobable.

You can verify this at home: some impromptu experimental philosophy on the floor of my study shows that it is easy, with minimal practice, to throw a die so that both in flight and after landing it executes a rolling motion that leaves the two faces on the axis of rotation almost no chance of ending uppermost. This, or a similar action, is characteristic of the "army blanket roll," a technique used by sharpers during and after the second world war to produce certain outcomes on the dice with statistics that defy the familiar uniform probability distribution.[3]

Where, then, does the uniform probability distribution come from? Again, it depends on the nature of the roll. When playing the craps in a casino, a die roll is required to hit the end wall of the table, which is covered in little rubber pyramids that effectively randomize the roll. (Some casino workers call this the "alligator wall.")[4] At home there are no alligator walls, but in the kind of roll described above, the die is shaken before it is thrown. It is the shaking that secures microconstancy, I will now show.[5]

In my paradigm case, the die-roller shakes the die heartily within the hollow of their clasped hands. You might with equal effect shake the die in a cup; either way, the die is thrown around inside a container of some sort, bouncing off the "walls" (perhaps walls of flesh) a number of times before it is released into the flight phase of the roll. Two dynamic factors contribute to the shaking: the somewhat irregular motion of the container, and the physics of the die as it bounces from wall to wall. Let me simplify things by assuming that the container does not move, and that the die's motion is entirely inertial, that is, that the die is released inside an unmoving container with some high initial velocity, thanks to which it rebounds from the walls a number of times before its eventual ejection. (Yes, like a gas molecule.)

The initial conditions of such a process are the die's initial position, translational velocity, and speed and axis of rotation (that is, angular velocity). The dynamics are microconstant with respect to some outcome of interest—for example, the die's having an axis of rotation piercing its 1 and 6 faces as it leaves the container—if the initial condition space can be divided into small regions in each of which the proportion of conditions producing the outcome is the same. I want to argue that the shaking process is microconstant in just this way.

Consider a situation in which a die is approaching one of the container's walls with a translational velocity, angular velocity, and position lying inside specified small ranges—thus, with a physical state lying in a small contiguous region of initial condition space S. Which initial conditions within S will lead to which outcomes? In particular, how will the die's axis of rotation after the collision depend on its exact state before the collision? The question can be visualized by graphing, either on paper or in your head, a set of evolution functions for the collision. Each such function will indicate, for some outcome of interest, such as the event of the post-collision rotation axis's falling within a certain range of possibilities, which initial conditions in S yield the outcome and which do not.

Almost all die-wall collisions consist in a corner of the die hitting the wall, and so the overall aspect of the corresponding evolution functions will be determined by the dynamics of corner-wall collisions. Typically, the corner of a die is slightly rounded, but whatever the exact shape, the outcome of a corner-wall collision will depend sensitively on the initial conditions: small changes in initial conditions will result in big differences in the post-collision state of the die.

Let me next run through the argument for microconstancy using an unrealistic supposition, which will then be relaxed. The unrealistic supposition is this: even when pre-collision initial conditions are constrained to a small region such as S, no post-collision axis of rotation is favored over any other. By this lack of favoritism or bias, I mean simply that, if you take a set of possible post-collision rotation axes and calculate the proportion of pre-collision initial conditions in S that produce an axis in the post-collision set, you will end up with pre-collision proportions that mirror the size of the post-collision sets.[6] (Note that the lack of bias is a purely dynamic property; probability has not yet been introduced to the picture.)

The microconstancy of the evolution function for a die roll's shaking phase, for any outcome of interest, follows almost immediately. Divide the space of initial conditions for the shaking phase into many small sets. Shaking, however it is done, will bring about several sequential collisions between die and wall. That the initial conditions of some particular episode fall into one of these sets constrains the parameters of these collisions, but not (if the choice of sets has been made wisely) any more than the assumption above that the collision parameters fall into a small region S. It follows that the dynamics of the collisions will spread the initial conditions within each small set evenly among post-collision axes of rotation. Choose an ultimate outcome, then—for example, the outcome that the die's axis of rotation when it enters the flight phase skewers the 1 and 6 faces—and the same proportion of initial conditions within each of the small sets will produce that outcome. Hence there is microconstancy, with a strike ratio proportional to the size of the set of rotation axes realizing the outcome, in this case one-third. With this microconstancy comes the randomizing power that, ultimately, equiprobabilifies a die roll's six outcomes.

That was straightforward, but only because I made things easy for myself. Let me see what I can do with a less idealized picture of the dynamics of collisions. Go back to the constrained set of pre-collision initial conditions S, representing a particular class of imminent die-wall collisions. As before, focus on one kind of post-collision outcome, the die's axis of rotation. I made two claims about the way in which the post-collision axis of rotation is determined by the pre-collision state. First, the axis of rotation is sensitive to small differences in initial conditions, even within S. Second, the conditions in S do not, given the dynamics, favor some axes over others.

It is easy to see that the first of these claims is correct; the second claim cannot be right, however, except perhaps for a rather narrow set of container and die geometries (and perhaps not even then). Suppose, then, more realistically, that for any particular small set of pre-collision states, there is a dynamic bias for some axes of rotation over others. Observe that globally, the set-by-set dynamic biases will cancel out, because of the symmetry of the die. To see this: consider a set of pre-collision conditions that favors a 1–6 axis of rotation over a 3–4 axis of rotation, in the sense that a greater proportion of states in the set produce a 1–6 axis than a 3–4 axis. Such a set has a mirror image such that, for each pre-collision state in the set, there is a state that is identical except that the die has been rotated 90 degrees around its 2–5 axis, switching the 1–6 and 3–4 axes. Because no relevant physical properties of the die are affected by the rotation,[7] this mirror image state favors the 3–4 axis of rotation over the 1–6 axis to precisely the same degree as the original state favored the latter over the former. And so on for every set of pre-collision states: hence, the biases cancel out, as claimed.

How does this help? Sensitivity to initial conditions is the key. To show you why, I am going to turn temporarily to an entirely different physical setup, a motley wheel of fortune, the relevance of which to die rolls will become clear soon enough.

The motley wheel is built like the regular wheel described in chapter 5, with the following three differences:

1. The red and black sections are large and unevenly distributed, as shown in figure 6.2: red dominates on some parts of the wheel and black on others.
2. Not a property of the wheel itself, but of its croupier: the maximum speed M that can be imparted to the wheel turns it only a single revolution.
3. Around the circumference of the wheel are inscribed possible spin speeds ranging from 0 to the maximum spin speed M, four times over, so that a complete range of possible speeds appears in each quarter of the wheel.

The evolution function for the wheel is shown in figure 6.3; as you can see, the wheel is not microconstant with respect to *red*. Rather, spin speeds at the high end of the allowed range are significantly more likely than low speeds to yield *red*.

Now suppose that the wheel is employed to carry out the following kind of operation. The wheel is spun. The outcome in the usual sense is ignored; instead, the spin is used to determine the speed for a second spin, by reading off the inscribed speed indicated by the pointer. The

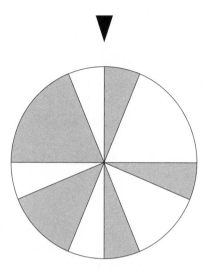

Figure 6.2: Motley wheel of fortune. Shaded sections are red, unshaded sections are black.

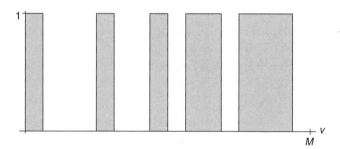

Figure 6.3: Evolution function for *red* on the motley wheel, for spin speeds ranging from 0 to the maximum possible spin speed M. The pointer is assumed to be at 12 o'clock, as shown in figure 6.2, and the wheel is spun anti-clockwise.

wheel is returned to its original position and spun with precisely the indicated speed. The resulting outcome, *red* or *black,* is noted. Call this a level one embedding trial.

The evolution function for the level one embedding trial is shown in figure 6.4. The reason for the term "embedding" should be clear: the new evolution function is constructed by compressing the evolution

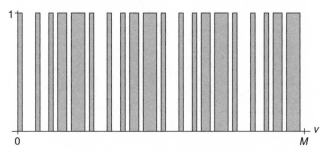

Figure 6.4: Level one embedding trial

function for a normal trial (figure 6.3) to a quarter of its original width and repeating it four times. This is a straightforward consequence of the way in which the speeds are inscribed on the wheel: as described above, the range of speeds is compressed and repeated four times around the wheel's circumference. The result is that, in an embedding trial, the first, speed-determining spin of the wheel maps the original spin speed v to a new speed of $4v$ modulo M.

Multiple levels of embedding are effected in the same way. In a level two embedding trial, the wheel is spun and used to determine a speed for a second trial, which is then used to determine a speed for a third trial. The outcome of the third trial determines the outcome of the whole, *red* or *black*. The evolution function for a level two embedding trial is shown in figure 6.5. It is, as you would expect given the recursive nature of embedding, constructed by compressing the evolution function for a level one trial (figure 6.4) to a quarter of its original width and repeating it four times, or equivalently, by compressing the original evolution function (figure 6.3) to 1/16 of its original width and repeating it 16 times. A level three embedding trial would compress the original evolution function to 1/64 of its original width and repeat it 64 times, and so on.

The level two dynamics is, as you can see from figure 6.5, microconstant with a strike ratio for *red* of one-half. The embedding process, then, takes a nonmicroconstant evolution function such as that in figure 6.3, and by compressing it and repeating it many times, creates a microconstant function with a strike ratio determined by the ratio of *red*-producing to *black*-producing initial conditions over the entire range of the original.

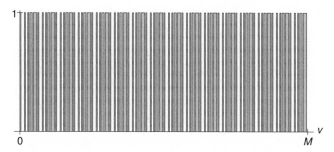

Figure 6.5: Level two embedding trial

To make microconstancy by embedding, two ingredients are needed. First, the process by which one trial creates the initial conditions for the next must blow up small differences, so that the initial conditions for later trials depend sensitively on those for earlier trials. The "blowing up" (or as I have called it elsewhere, *inflation*) need not be extreme; in the motley wheel it is merely a four-fold linear expansion. Multiple embeddings compound the expansion exponentially, providing more than enough compressing power to manufacture microconstancy.

Second, the inflation must be sufficiently uniform that it creates identical copies of the original evolution function, or at least, copies with the same strike ratio for the outcome of interest. In the motley wheel of fortune, this is achieved by inscribing the speeds in the same way in each of the wheel's quadrants. Had some quadrants emphasized low speeds— had the inscription in these quadrants spread out the low values and compressed the high values—and other quadrants high speeds, the ratio of *red*-producing speeds would not be constant across the evolution functions in figures 6.4 and 6.5.[8]

Return now to the dynamics of die-shaking. That a shaken die collides several times with its container's walls creates the same embedding effect as seen with the motley wheel. The initial conditions of the second such collision will depend on the first, and they will do so sensitively. Further, because of the symmetry of the die (and the smooth geometry of the container), it seems likely that the form of the dependence is much the same whatever route the die takes around the container in the course of the shaking. As a consequence, the evolution function for the shaking process, relative to any outcome of interest, will be made up of many compressed copies (or near-enough copies) of the evolution function for a shaking process in which only one collision occurs.[9]

I argued above that, because of the symmetry of the die, the evolution function for the one-collision shaking process favors overall no axis of rotation over any other. The evolution function for a shaking process with several collisions, then, will be made up of many small regions in which equal numbers of initial conditions lead to each of the three possible axes of rotation; it is therefore microconstant, with a one-third strike ratio for each axis of rotation. The same goes for other outcomes of interest, such as the identity of the uppermost face as the die leaves the container—the strike ratio for each face in this case being one-sixth.

Shaking, then, is an effective randomizer, if it goes on long enough to achieve the sort of embedding shown in figure 6.5: the shaking process is microconstant with respect to all the die's properties of interest as it leaves the hands or container, with strike ratios that mirror the die's symmetries. The other two phases of a die roll, the die's spinning through the air and then tumbling on the landing surface, need only preserve this microconstancy, as the tossed coin's bounce preserves the microconstancy of the tossing process (section 6.1), for the die roll process as a whole to be microconstant with equal strike ratios for each of the six outcomes. Given a microequiprobabilistic initial condition distribution, the familiar uniform physical probability distribution over the outcomes is explained. Given a microequiprobabilistic joint distribution (the distribution over initial conditions of sets of die rolls), the stochastic independence of the outcomes—hence their Bernoullian behavior—is also explained (section 5.3).[10]

Most important to the explanation are two properties of the die's dynamics, namely, its sensitivity to initial conditions and its symmetry, which dynamical symmetry of course depends in turn on the physical symmetry of the object—the die—itself. Sensitive dependence is responsible for embedding; symmetry is responsible for the uniformity of the embedding, so that the same strike ratios are found in each part of the evolution function for shaking, and also for those strike ratios being equal for the critical outcomes, such as the three axes of rotation and the six faces.

6.3 Brain and Die

We look at a tumbling die and "see" that the probability of rolling a 5 is one-sixth. What is the source of this insight?

One possible explanation attributes to us—both adults and young children, if the latter turn out to be as savvy about dice as they are about urns—the reasoning laid out in the previous section. On this hypothesis, then, we infer the microconstancy of the evolution function for a die roll by perceiving that the sensitive, symmetrical dynamics induced by collisions when a die is shaken has an embedding aspect, and that the embedded pattern will favor each outcome equally.

Given the difficulty of the derivation, however—complex enough even in what was a mere sketch of the outline of a proof above—this proposal would seem to take an awfully optimistic line on the power of human physical intuition. Certainly, we do not consciously reason our way through the intricate dynamics of the evolution function. Is there some behind-the-scenes homunculus quietly doing the math without bothering central cognition? The existence of an inner physics prodigy who calculates probabilities on our behalf without displaying his working cannot be ruled out; indeed, we know that such characters are vital to the functioning of the visual and other perceptual systems. But it is surely worth considering some alternative hypotheses as to the form of our equidynamic reasoning about dice.

At the end of chapter 5, I suggested a simple heuristic for inferring a uniform physical probability distribution over the outcomes of a physical process: if the outcomes are sensitive to initial conditions, and if the apparatus that produces the outcomes is symmetrical in the relevant ways—with the question of which ways are relevant temporarily deferred—then infer a uniform distribution. (Infer independence of the outcomes, too.) Provided that sensitivity and symmetry are typically found in contexts where the other conditions required for embedding (or stirring, or whatever) hold, then the inference rule, though not infallible, may be highly reliable.

This simple story does not, however, capture all that is going on in equidynamic reasoning. We do not blindly infer uniformity of probability whenever we see sensitivity and symmetry. Consider, for example, the "army blanket roll" described in the previous section, in which a die is rolled along a blanket or rug so that the faces pierced by the die's axis of rotation have little or no chance of landing uppermost. An attentive, intelligent witness to such a roll (though not, evidently, its historical victims) can see immediately that the process is not equally likely to yield all six outcomes. Something about the physics of the roll deters us from making the inference to a uniform probability distribution.

What might that be? In answering this question, I hope to find a hypothesis about our dice-related inferences that hits the golden mean, lying partway between the simplistic sensitivity-and-symmetry account, which fails to explain our inferential sophistication, and the derivation-of-microconstancy-from-physics account, which attributes to our brains what is perhaps an implausible mathematical prowess.

Let me, then, consider three possible accounts of our reasons for concluding that the army blanket roll's outcomes are not equally probable. First, even if we take a simple sensitivity-symmetry approach, the inference to equiprobability for the army blanket roll might be defeated by information about statistical patterns in the outcomes. If we come to know that some faces are landing uppermost much less often than others, we have good reason to think that the conditions under which sensitivity and symmetry induce a uniform probability distribution are not all present. Statistical defeat is real and important, but it cannot explain our reluctance to infer uniformity in the army blanket case, as we seem to know in advance that the statistics will not come out right—we predict nonuniform statistics on the basis of our knowledge of the nonuniformity of the probabilities, rather than inferring skewed probabilities from skewed statistics.

Second, once the requirement of symmetry in our rule for inferring equiprobability is spelled out in greater detail, it might turn out that the army blanket roll is not sufficiently symmetrical to license an inference to a uniform probability distribution. What is the symmetry-breaker? Is it the setting of the die, that is, the deliberate and nonrandom orientating of the die before the roll? But setting does not always dissuade us from inferring uniform probabilities: if between the setting and the rolling, the die is vigorously shaken, we are happy to infer uniformity. Thus, it is not easy to pin the difference between these processes on the symmetries of the parts alone.

My third and final suggestion: we resist the inference to equiprobability because we know that, as the die rolls along the blanket, some faces for systematic reasons spend far less time uppermost than others. More generally, the inference to uniformity is resisted if the device at the center of the process—in this case, the die—fails, in the course of the process, to visit the different relevant states with equal assiduousness. Equal visiting time might be a precondition for the inference to uniformity, or unequal visiting time might be a defeater. Let me lay out a version of the "precondition" approach, on which some sort of premise about equal visitation rates plays an essential role in our inferring that a die roll's outcomes are equiprobable.

You have seen a visitation rate story already; chapter 5's stirring processes have microconstant evolution functions, hence outcome probabilities equal to strike ratios, because outcomes are determined by a stirring parameter that cycles through or "visits" its possible values in a regular way.

A die roll does not exhibit stirring behavior: the only point at which it cycles regularly through values is in its flight phase (and in an army blanket roll or similar, its bouncing phase), but in the flight phase not all potential outcomes are visited equally. What I want to talk about is "equal visitation" behavior in the shaking stage. To this end, identify a property of the shaken die that will, once the die is released, help to determine the outcome of the roll. Reasonable choices include the die's axis of rotation, its uppermost face, and so on. Call this a *shaking parameter*. When the die is flying around in your cupped hands, the shaking parameter makes something of a tour of its possible values, but there is no predictable progression to this tour; rather, the twists and turns of its itinerary are quite irregular. Intuitively, however, the irregularity is egalitarian in the "equal opportunity" sense: it does not inherently favor any particular value of the parameter over any other. It is this equal opportunity visitation that, I suggest, motivates our ascription of equal probabilities to the outcomes of the die roll.

There are two ways you might make sense of the metaphor of equal opportunity. On the one hand, you might think in terms of a probabilistic dynamics: you might say that the shaking parameter has an equal probability of passing through any of its values in the course of the shaking, or (not quite the same thing) that it takes a "random walk" through the space of its possible values. In so doing, you appear to attribute a stochastic microdynamics to the die's motions.

On the other hand, you might think in terms of the microconstancy of a more or less deterministic dynamics: you might say that within any small set of initial conditions, there are just as many that lead, after a fixed time, to the shaking parameter's having one value as there are leading to its having any other. (To put this more carefully: the initial condition space can be divided into small contiguous sets within each of which, for any two equal-sized intervals of the shaking parameter's range, equal proportions of initial conditions lead to the parameter's falling after a fixed time into each of those intervals.) The evolution function for the event of the shaking parameter's falling into a given set after a given period of time is then microconstant with a strike ratio proportional to the size of the set—which in conjunction with microequiprobabilistic initial conditions implies a uniform probability distribution.

The embedding aspect of die dynamics provides a physical foundation for both kinds of thinking. (The connection between embedding and microconstancy was explained above; for the connection between embedding and the stochastic microdynamics approach, you will have to wait for the next chapter.) The microconstancy conception provides a clearer view of the foundation, but it is the "random walk" conception that figures in our equidynamic reasoning about the dice and other stochastic processes, I propose: we understand the die's bouncing around inside our hands as having a stochastic microdynamics, its shaking parameter visiting various possible values at random. The pattern of its visits to these values is therefore irregular, but the probability distribution over the visits is quite regular—uniform, in fact, or near enough to uniform to ensure the equiprobability of the ultimate outcomes. It is the processes meriting this treatment that I call the *shaking processes*.[11]

At this point I will leave my dealings with dice suspended in midflight. If you do not see clearly what a shaking process is, or what kind of stochastic microdynamics we attribute, on my view, to the dice, or how we draw conclusions about visitation probabilities from the microdynamics, you have not missed anything: I have not yet given a substantive account of any of these matters. The discussion resumes in chapter 7, where I give a general account of equidynamic thinking about random walks, hence about "shaking processes," applicable to dice, bouncing balls, and more; in section 8.1, I then apply this account specifically to the case of the dice.

6.4 Urn Studies

Three questions about drawings of colored balls from an urn: Where does the Bernoulli distribution over outcomes come from? How do adults infer this probability distribution? And something I was not in a position to ask about coins or dice: does the evidence from developmental psychology suggest that, in their thinking about urn probabilities, infants differ from adults?

First, the probability itself. As with a die roll, there are qualitatively different ways to draw a ball from an urn. To see this, divide a drawing into two phases: the shuffling and the drawing. In the shuffling phase, the balls are mixed in the urn. One way to shuffle the balls is to give the urn a good shake; another is to pour the balls into the urn from another container in a sufficiently rough and randomizing manner. In the drawing phase the croupier reaches into the urn, perhaps rummaging around, and pulls out a ball. The importance, to randomization, of the rummaging during drawing depends on the thoroughness of the shuffling: the more shuffling, the less rummaging is required.

(Another way that drawings may differ: when drawing more than one ball, the balls previously selected may either be replaced or retained. With replacement, the same Bernoulli distribution characterizes each draw; without replacement, the relevant Bernoulli distribution changes.)

Consider a drawing in which the urn is given a good shake before each selection, as in Xu and Garcia (2008)'s experimental paradigm. To keep things relatively simple, suppose that there is no rummaging whatsoever: the croupier's hand moves to the same position every time and selects the nearest ball (in some fixed sense of "nearest"); furthermore, although this reaching may push some balls out of the way, it does so in the same way on each drawing. The probabilistic aspect of such a drawing therefore depends entirely on the power of the shaking to effect a shuffling.

How does shaking achieve this end? Call the region in the urn from which the ball will be drawn the "target zone." The outcome of a drawing is decided by the color of the ball that is passing through the target zone at the time the shaking stops, so the shuffling power of the shaking in these circumstances subsists in its ability to send what is intuitively a random stream of balls through the target zone.

To analyze this complex process, take an arbitrary ball and examine the evolution function for the event of the ball's occupying the target zone when the urn is done shaking. If the evolution function is microconstant, you have much of what you need to understand the Bernoulli distribution over urn-drawings. (The rest will hinge on the strike ratios, concerning which more shortly.)

Let me make a prima facie argument for microconstancy by drawing a physical analogy between the shaken ball and the tumbling die. Like the die, the ball experiences a series of state-changing collisions. Some of these send it toward the target zone; some send it in other directions. The dynamics of collisions between shaken balls share two properties with the dynamics of collisions between a shaken die and its container walls: post-collision states depend sensitively on pre-collision states, and the overall form of the dependence is qualitatively the same for any collision.

As a result, the dynamics of ball-ball collisions is, like the dynamics of die-wall collisions, embedding: the evolution function for a path containing $n+1$ collisions is made up of many miniaturized copies of the evolution function for an n-collision path. (If the miniatures are not literally copies, they are at least qualitatively similar to one another, so that they have equal strike ratios for the outcome of interest—though as remarked in note 8, a weaker condition still would suffice.) With sufficiently many collisions, the embedding creates microconstancy. This line of thought is, needless to say, far from rigorous. I hope it is plausible enough, however, to go on with for now.

Suppose, then, that the evolution function for a shaken ball's ending up in the target zone is microconstant. What is the strike ratio for this outcome? The question need not be answered. To secure the sought-after Bernoulli distribution, it is enough that the strike ratio is the same for any ball in the container. If the container is shaken for long enough, it seems plausible that the strike ratios for physically identical balls might be equal: all that distinguishes such balls are their positions when the shaking begins, and if the balls move around the container for long enough, their starting positions hardly matter.

But most actual urn-shakings do not, I suspect, go on for long enough to have this effect. After a standard short, sharp shaking, a ball that is far from the target zone will be less likely to have reached the zone than a ball that is close, or to put things in purely dynamical terms, a greater proportion of the initial conditions will lead to the target zone for balls close to the zone than for those far from the zone. (Gravity will also play a role, but in what follows, let me for simplicity's sake ignore it to focus on distance.)

If this is correct, then not every starting configuration for the balls in the urn will, after an abbreviated shake, produce the expected probabilities: if, for example, half the balls in the urn are red and half white, but the majority of balls close to the target zone are red, then the probability of *red* will be higher than one-half (though the distribution is nevertheless Bernoullian).

Further, even when the probability of *red* after a short shaking is one-half, it will not be because any ball is equally likely to be drawn, but because in the pre-shaking state, there is no correlation between color and distance to the target zone, that is, because for any given distance, the balls sitting at approximately that distance from the target zone are about one-half red and one-half white—or if there are imbalances in the groups at different distances, because those imbalances cancel out overall.

In order for the probability of *red* to equal the proportion of red balls in the urn, then, the duration of the shaking should be of a length determined by the prior organization of the balls: a brief shake is enough when the colors are evenly distributed, but a longer period of shaking is needed when the distribution is uneven.

Supposing that my surmises about the microconstancy of shaken-urn dynamics are correct, then the urn, like the other gambling setups discussed so far, offers an epistemic affordance: there is a strong relationship between the probabilities of drawing balls of various colors from the urn and the physical properties of the urn. Most obviously, the probability of, say, *red* is equal, when the circumstances are right, to the proportion of red balls in the urn. (I assume, though I will not prolong the exposition by trying to establish, that the same can be said for drawings in which there is a substantive rummaging process, rather than a fixed target zone.)

The simplest inference rule we might use to take advantage of this fact—or at least, the simplest rule that makes no essential reference to urns and balls—is the sensitivity-symmetry rule described earlier (sections 5.4 and 6.1): when you have sensitivity to initial conditions and relevant physical symmetries, infer equal physical probabilities. In this case, the equal probabilities are of any ball's being chosen; taking account of the balls' colors then gives you a probability for each color proportional to that color's representation in the population.

Is our reasoning about urns so uncomplicated? As in the case of the die, I think not; there is strong evidence that we take into account aspects of the process producing the outcome that go beyond sensitivity and symmetry. Consider, in particular, the observation made just a few paragraphs ago: when the colors are unevenly distributed in the urn, a long shaking is needed to produce outcomes with probabilities proportional to color, while when the colors are evenly distributed, a short shaking will suffice. This fact is, I suggest, known to the average amateur urn theorist. That is, a normal human, unacquainted with anything like the microconstancy-based treatment of urn probabilities above, will recognize that the relationship between initial color distribution, shaking period, and probabilities holds. More revealing still, they will see that when the distribution is uneven and the shaking period is short, a draw of the color that is overrepresented near the target zone will be more probable than its overall representation in the population suggests.

Such insights require the thinker to consider the way that shaking moves individual balls around the box. I suggest that the relevant reasoning proceeds, in adult humans, in much the same way as in the case of the shaken die: like the die and other objects involved in shaking processes, the balls in a shaken urn are conceived of as taking a random walk through the space of possible states. (You might represent the system using a set of shaking parameters, one for each ball, or you might use a single shaking parameter for the whole setup such as the color of the ball in the target zone at any given time. Either way, you think in terms of the trajectories of individual balls.) As noted in the previous section, our equidynamic reasoning about random walks in general will be examined at length in chapter 7; the resulting theory is then applied to urns in section 8.2.

The case of urn-drawings is the first to be considered in which there is direct evidence about the nature of infants' equidynamic reasoning. Do infants think about urns in the same way as adults? Perhaps they think about them in exactly the same way. Perhaps they have a more rudimentary set of inference rules that are fleshed out as they learn more about the world. Or perhaps they think about urns using a rule quite unlike the adult rule.

Let me consider two more concrete possibilities. On what you might call the dynamic hypothesis, infants' thinking resembles adults' in that, like adults, they require sensitivity of outcomes to initial conditions and the presence of certain dynamically relevant physical properties, including physical symmetries, as necessary conditions for inferring physical probabilities. (The dynamic hypothesis allows, but does not require, that infants' dynamic reasoning is identical to adults'; it is agnostic, then, as to whether infants think explicitly in terms of random walks and so on.)

On what you might call the nondynamic hypothesis, infants pay little or no attention to the physical dynamics by which outcomes are determined; they base their inferences of physical probabilities (or some surrogate for physical probabilities) on physical properties, but they do not think of those properties as determiners of dynamics. In the case of the urn, for example, what might be decisive for infants is the fact that the arrangement of balls in the urn—or rather, in the transparent box—is haphazard or random (in the static sense of these words). They consider any selection from such a population equiprobable.

The nondynamic theory would have to be augmented in several ways to account for the findings, reported in chapter 4, of Xu and Garcia (2008), Xu and Denison (2009) and Denison and Xu (2010a).

First, when infants are shown the contents of the box and then tested for their expectations concerning a small sample, the box is covered and shaken after they see the haphazard arrangement of balls and before the sample is drawn. They do not have direct observational knowledge, then, of the haphazardness of the arrangement from which the balls are sampled, but they infer probabilities anyway. To account for this fact, the nondynamic theory might posit that infants assume that shaking preserves haphazardness.

Second, infants do not assign the usual probabilities when they believe that a sample is selected deliberately; they expect to see not the most populous color but the sampler's preferred color (Xu and Denison 2009). The nondynamic theory might account for this fact by assuming that preference-based reasoning about intentional agents trumps structure-based reasoning about physical probabilities. According to this view, structure-to-probability reasoning does not rule itself out of contention in a case of deliberate sampling (whereas on the dynamic theory, it does, since deliberate sampling is not sensitive to initial conditions); it is rather overruled from without.

Third, infants do not figure into their probability calculations balls that are stuck to the box and so cannot be removed. If a box contains 50 percent immovable green balls, 40 percent movable red balls, and 10 percent movable white balls, they expect a five-ball sample to contain about four red balls, one white ball, and no green balls (Denison and Xu 2010a). Surely some minimally dynamical premise must be attributed to infants to explain this inference.

Once the nondynamic view is saddled with these additional assumptions, it loses its luster, I suggest. The dynamic view (or rather, family of views) is more unified and no more complex; that it implies a continuity between infant and adult structure-to-probability inference is especially welcome. Infants, just as much as adults, are equidynamic reasoners in the full sense of the term.

Is it plausible, though, to suppose that six-month-olds are capable of seeing that the outcome of a drawing from an urn is sensitive to initial conditions? It is hard to say. We know that infants have far more sophisticated physical intuitions than was once supposed (Mehler and Dupoux 1994; Sperber et al. 1995; Carey 2009). We know also that some sort of dynamic reasoning is going on in urn scenarios, in the light of the case of the sticky balls. But that is all. Further experimental results are needed; what would infants think, for example, about a box of balls in which all white balls were at the bottom corner, and from which a ball was drawn without shaking?

7

BOUNCING

Adults, like the children tested by Téglás et al. (2007), are at home probabilistically with bouncing balls, or so I suggested in chapter 4: we look at a collection of equally sized balls or other objects careening around a container and infer that each is equally likely, after a given time period, to exit the container, or that such an exit is equally likely to occur by way of any one of multiple equally sized apertures. What rules guide our equidynamic reasoning about bouncing?

There are two broad classes of equidynamic rules of inference that we might apply to the problem: short-term and long-term rules.

A *long-term* inference rule ascribes probabilities to properties of a bouncing ball, or other such object—to its position, its direction of travel, and its speed—only after a certain "relaxation period" has elapsed. Intuitively, the ball's initial position and velocity, and the initial state of the system as a whole, are rendered irrelevant by a process of randomization that goes on throughout the relaxation period; once the period is over, the influence of the initial conditions is entirely washed away, leaving a probability distribution determined by dynamics alone. In this chapter I will sketch a pair of long-term rules warranting, under certain circumstances—most importantly the presence of an appropriate randomizing dynamics—the inference of a uniform distribution over a bouncing ball's position and direction of travel.

A *short-term* inference rule allows you to infer a probability distribution over properties of a bouncing ball without waiting for relaxation. My paradigm is Maxwell's microequiprobability assumption: conditional on a molecule's being in a certain small region, it is equally likely to be anywhere in that region, or in other words, the probability distribution over molecular position is at all times approximately uniform over small areas. Maxwell's "unofficial" argument, you will recall, uses this assumption to infer a trend toward a certain equilibrium distribution, rather than treating it merely as a property of the equilibrium distribution (section 2.3). Thus, the unofficial argument treats the assumption as valid in the short term as well as in the long term.

In the first section of this chapter, I transmute the assumption of position microequiprobability into a full-fledged equidynamic rule by spelling out a set of conditions under which we humans are disposed (or so I propose) to infer microequiprobability. It is the short-term microequiprobability derived from such a rule that underwrites the randomization premise required for the application of the long-term rules that take you to the uniform distribution, post-relaxation, of a bouncing ball's position and direction of travel. Anyone can follow in Maxwell's footsteps, then, provided they hold tight to the equidynamic handrail.

7.1 The Microdynamic Rule

No preliminaries; I hereby submit the following short-term inference rule for your consideration. Take an entity—a ball or other bouncing object, for example—whose trajectory is determined by a series of interactions with other objects, moving or stationary (balls, walls, floors, whatever). The entity's state at any time can be characterized by a set of variables, such as position and velocity, that I will call simply its *state variables*. Suppose that each of the state-determining interactions (there may be only a single kind, or there may be more than one) satisfy the following two conditions:

Sensitivity: The entity's post-interaction state—in the case of the ball, its position and velocity—is highly sensitive to the parameters of the interaction. In other words, the entity's post-interaction state variables depend sensitively on its pre-interaction state variables.

Smoothness: The dynamics of the interactions are "smooth" in the sense that they have a property called microlinearity, to be defined below.

Then you may infer a microequiprobabilistic probability distribution over each of the state variables of any such entity.

This I call the *microdynamic rule;* I hypothesize a commitment to the rule—or something like it—in all normal humans, certainly adults and possibly children too.

The distribution imputed by the microdynamic rule is guaranteed to have three independence properties. First, its microequiprobabilistic aspect is independent of the concurrent values of the entity's other state variables, which is to say that when the microdynamic rule applies, the distribution over any small interval of one of an entity's state variables, conditional on the values of the entity's other state variables, is uniform. For example, the microdynamic rule allows you to infer of a bouncing ball that the distribution over its position within any small region—what you might call the distribution over its "microposition"—is uniform regardless of the ball's velocity.

Second, the distribution's microequiprobability is independent in the same way of the state variables of other entities in the system. Thus, the microposition of one ball bouncing in a box is uniformly distributed given any assignment of positions and velocities to the other balls in the box. (It is quite natural, I think, to understand the claim of microequiprobability as concerning the conditional distributions, and so as incorporating implicitly these first two independence properties.)

The third independence property is a generalization of a property you have already encountered in bouncing balls, namely, that the post-collision microposition of a ball is independent of its pre-collision microposition. Suppose that the value of a state variable falls into a microsized interval I_1 before an interaction and a microsized interval I_2 after the interaction. Then the position of the variable within I_2—the analog of post-interaction microposition—is stochastically independent of its position within I_1. The same is also true for two different state variables of the same entity, and by extension, for state variables of two different entities.[1]

I will say nothing further about the microdynamic rule's requirement of sensitivity to initial conditions, except to note that collisions between spheres and similar objects satisfy the requirement. (Collisions between spheres and a straight wall do not satisfy the requirement, I think; the occurrence of intersphere collisions is necessary, then, to apply the rule to balls bouncing in a box. Collisions between dice and a straight wall, by contrast, do satisfy the requirement, for the reason noted in section 6.2.)

The microdynamic rule's smoothness component requires the dynamics of interactions among entities to have the property of *microlinearity*. A process mapping pre-interaction conditions to post-interaction conditions has a microlinear dynamics if it is approximately linear over any contiguous, microsized region of the initial conditions. The process as a whole can be approximated, in that case, by a patchwork of linear functions over microsized domains.

The microlinearity of a dynamics is relative to two things: a specification of what is small enough to be "micro," and a means of quantifying the physical magnitudes—position, velocity, and so on—upon which the dynamics acts. Let me deal with these relativities in turn.

First, the specification of "micro" need not be fixed in advance; rather, the degree of "micro" relative to which microlinearity obtains will, along with the degree of initial condition sensitivity, determine the degree of "micro" relative to which microequiprobability may be inferred using the microdynamic rule, as will become clear shortly.

Second, a means of quantifying the relevant physical magnitudes is, I propose, built into the microdynamic rule: to satisfy the smoothness requirement, the dynamics' microlinearity must be with respect to a set of standard variables (sections 5.6 and 12.1). Which standard quantification to use? For velocity, should you use polar or Cartesian coordinates? The question is moot, because microlinearity with respect to one standard variable implies microlinearity with respect to any other standard quantification of the same physical property (this because, as explained in section 12.3, alternative standard quantifications of a property, where they exist, are microlinear functions of one another), so a dynamics that satisfies the smoothness requirement with respect to one set of standard variables satisfies it also with respect to any other.

Finally, it may help to give an example of a dynamics that lacks microlinearity. Imagine a ball sitting in a depression with two hollows. The ball is occasionally pushed around by impacts from outside, but it always settles back to one of the two hollows. Which hollow it settles into depends sensitively on the preceding impact. But the distribution over post-impact conditions is not microequiprobabilistic; in fact, only two precise endpoints are possible. Such a dynamics is not microlinear, and so fails the smoothness test.

We humans evidently need no justification to prompt our application of the microdynamic rule in both science and everyday life; our allegiance to equidynamic thinking has deeper roots than mere philosophical argumentation. But as a reader of this book, you are surely asking in your most philosophical tone of voice: Is the rule reliable? If so, why? (The forthcoming justification of the microdynamic rule is not essential for a grasp of anything beyond this section; it is therefore safe to move on to section 7.2.)

The microdynamic rule is not guaranteed to yield true conclusions, since satisfying the rule's sensitivity and smoothness requirements is not a logically sufficient condition for the conclusions about microequiprobability that the rule purports to warrant. But in normal conditions, it nevertheless works well enough. Let me explain why.

Suppose that the sensitivity and smoothness tests are applied to an object whose state variables are determined by many iterations of a single kind of interaction, such as inter-ball collisions. (I will leave generalization as an exercise for the reader.) Here are two conditions that are jointly sufficient for the microdynamic rule's reliability, meaning that if the dynamics of the interaction passes the sensitivity and smoothness tests and these conditions hold, then the object's state variables are sure to be microequiprobabilistically distributed:

1. If an interaction has a dynamics that is microlinear and sensitive to initial conditions, then a microequiprobabilistic distribution over its pre-interaction conditions will induce a microequiprobabilistic distribution over its post-interaction conditions.
2. Pre-interaction conditions tend to be microequiprobabilistically distributed.

I will demonstrate that (1) is true and then suggest that (2) typically holds, showing you that its holding does not render the test superfluous.

Condition (1) first. The task is to show, informally, that a microlinear dynamics that is sensitive to initial conditions—or as I sometimes put it to save words, a dynamics that is *inflationary*—maps a microequiprobabilistic distribution over pre-interaction conditions to a microequiprobabilistic distribution over post-interaction conditions.

A linear function from a set of pre-interaction conditions to a set of post-interaction conditions has the following notable property: a uniform probability distribution over the initial conditions is mapped to a uniform probability distribution over the post-interaction conditions.

Divide the space of pre-interaction conditions for an inflationary microlinear transformation into contiguous microsized sets. Given microequiprobability, the probability distribution over each such set is uniform. Given microlinearity, the effect of the dynamics on each such set is approximately that of some linear function. It follows that the distribution over the image of such a set (the set of post-interaction conditions onto which the dynamics maps the set's pre-interaction conditions) is also uniform, and the image is contiguous. Given inflation, the image is much bigger than the original set, thus much bigger than microsized—it is macrosized. The distribution over the post-interaction conditions is therefore a weighted sum of uniform distributions over contiguous macrosized subsets of the space. Such a distribution will be approximately uniform over most microsized regions, hence microequiprobabilistic. Thus, an inflationary microlinear dynamics maps microequiprobability to microequiprobability. (The independence component of microequiprobability will be discussed shortly.)

The above justification vindicates the conclusion that the joint distribution over the state variables of all entities involved in the interaction is microequiprobabilistic. What was wanted is a microequiprobabilistic distribution over each state variable, conditional on the values of the other variables. This follows immediately from the microequiprobability of the joint density.[2]

The other component of microequiprobability is independence: the microdynamic rule guarantees, for example, that a bouncing ball's post-collision microposition is stochastically independent of its pre-collision microposition. How can that be? A perplexing question; the pre-interaction microposition helps, after all, to determine the post-interaction microposition.

The treatment of this sort of independence in Strevens (2003) suggests the following answer. The stochastic independence relation does not, in fact, hold. But something else does hold that is for practical purposes as good as independence, namely, the independence of *approximate* post-interaction microposition from *approximate* pre-interaction microposition. (It is as good for practical purposes because, in contexts where the independence property is put to work, independence of approximate position is typically sufficient for the job at hand.)

To understand what is going on here, three levels of information about position have to be distinguished: high-level information, which determines into which microsized region a ball's position falls; medium-level information, which determines in which subregion of the microsized region it falls; and low-level information, which determines exactly where in this subregion it falls. The post-interaction mid-level information is independent of the pre-interaction mid-level information, and this is sufficient, in almost any context where the independence component of microequiprobability is put to work, to secure what is wanted. I will not try to elucidate here. Strevens (2003) defines the relevant notion of levels of information (section 2.7), defines what it means for levels of information to be independent (section 3.4), and shows that an inflationary microlinear dynamics is sufficient for independence of the mid-level information (section 3.7, section 4.8). I hope that this rough outline provides you with some sense of the important issues and moves to be made.

Finally, the sine qua non: condition (2). That a dynamics transforms a microequiprobabilistic distribution over pre-interaction conditions into a microequiprobabilistic distribution over post-interaction conditions is of slim interest unless the pre-interaction distribution is in fact microequiprobabilistic. Thus, it would seem, on the approach to justifying the microdynamic rule offered here, we need to have some grounds for imputing microequiprobability to an interaction's initial conditions.

But if we have such grounds, what contribution could the microdynamic rule possibly make? What is the use of a rule licensing an assumption of microequiprobability for a set of state variables, if its application requires an assumption of microequiprobability for the very same variables? What can a dynamics' inflationary microlinearity tell you that you did not already know?

Microequiprobability, you will recall, comes in degrees. A distribution is microequiprobabilistic if it is approximately uniform over any micro-sized region; the import of the definition is, then, relative to a standard for what is "micro." The more liberal this standard—the larger the regions that count as microsized—the stronger the resulting definition of microequiprobability, since uniformity over a large region is harder to come by than uniformity over a small region. In a slogan: the larger the "micro," the stronger the microequiprobability.

Here is what inflationary microlinearity can do for you: given only the weak microequiprobability of the pre-interaction conditions, it delivers strong microequiprobability of the post-interaction conditions. It does not manufacture microequiprobability from scratch, then, but it strengthens it wherever it finds it. Repeated applications of microlinear inflation strengthen microequiprobability many times over (up to a maximum strength determined by the dynamics' sensitivity to initial conditions and the standard of "micro" relative to which it is microlinear). Inflationary microlinearity is, in short, a microequiprobability purifier and amplifier.

Throw a little low-grade microequiprobability into a system with an inflationary microlinear dynamics, and you get a lot of high-grade microequiprobability shortly thereafter.

The success of the microdynamic rule does rest, then, on an assumption of microequiprobability, but what is required is microequiprobability in a much weaker form than can be inferred from the rule.

To sum up: Suppose that it is true of the typical system to which we apply equidynamic reasoning that the system's state variables have at least a low-grade microequiprobability at any time, perhaps introduced by environmental jangling (section 12.3). Then in those systems that satisfy the microdynamic rule's sensitivity and smoothness conditions, a strong, high-grade microequiprobability is continually being generated over the state variables. The microdynamic rule is, consequently, ecologically valid.[3]

7.2 The Equilibrium Rule

You have used the microdynamic rule, let me suppose, to infer a microequiprobabilistic distribution over the positions of the balls bouncing around a certain box. What then? You may do what Maxwell did: thinking of microposition as an initial condition, combine the microequiprobability of position with facts about collision dynamics to infer a probability distribution over the outcomes of collisions. This gives you a short-term microdynamics of collision that is in part probabilistic. And then? Again following Maxwell, you might use the probabilistic microdynamics to draw conclusions about the longer-term distribution of the balls.

One arduous way forward is to build a long-term probabilistic dynamics from the short-term dynamics by hand: begin with some distribution of the balls' positions and velocities; apply the microdynamics to those balls that collide; derive a post-collision probability distribution over the positions and velocities; repeat. Even thinking through the process in the abstract is tiring. In the first three propositions of his 1859 derivation of the velocity distribution, Maxwell looks to be heading in this direction, but after a few sentences he breaks off and concludes in the final paragraph of proposition III that, whatever the details, "after a certain time the [kinetic energy] will be divided among the particles according to some regular [statistical] law." (The passage is reproduced and discussed in section 2.3.)

This, I will now suggest, is unadulterated equidynamics: there is an inference rule available to us all, paradigm-shifters and ordinary citizens alike, warranting just such a conclusion, that aspects of a system satisfying certain conditions will settle down to a fixed and unique distribution—a global stable equilibrium. My hypothesis as to the form and the grounds of the rule, which I will call the *equilibrium rule*, are discussed in this section.

A further rule, the *uniformity rule*, applies to a subset of the variables for which the existence of an equilibrium distribution can, by way of the equilibrium rule, be derived. It warrants the conclusion that the equilibrium distribution over the variables to which it applies is uniform. It can also be used to infer systematic deviations from uniformity in some equilibrium distributions. The uniformity rule is stated, and its reliability discussed, in section 7.3.

The form of the equilibrium rule is, I hypothesize, as follows. Suppose you have a system containing one or more entities, perhaps of different kinds, with the state of each kind of entity characterized by a set of state variables, such as, in the case of a bouncing ball, position and velocity. Then you may conclude that the probability distribution over a state variable for a given kind of entity has settled down to a unique equilibrium if the following conditions hold:

> *Randomization:* Changes of the state variable in question depend
> sensitively on a stochastic variable (defined below).
> *Boundedness:* The values of the state variable fall within a finite range;
> they cannot go to infinity.
> *Relaxation:* The system has been "running" for a sufficiently long time.

It is not clear to me whether the equilibrium rule also licenses the conclusion that the state of any one entity with respect to the variable in question is independent of the states of any other entities. (I think that it does not license the conclusion that the state variables of a single entity are independent of one another, for example, that a bouncing ball's position, even if it passes the test, is independent of its velocity.)

A variable is "stochastic" if (a) there is a (nontrivial) physical probability distribution over the full range of the variable's possible values, (b) this distribution is independent of all of the system's state variables (or at least, independent of their approximate values), in the sense that conditionalizing on the (approximate) values of the state variables at any time makes no difference to the form of the distribution, and (c) values of the variable at points sufficiently separated in time and space are stochastically independent. The equilibrium rule's randomization requirement is satisfied, then, if there exists a variable satisfying these three conditions and on which changes of the state variable in question sensitively depend.

In the case of the bouncing balls, microposition of course counts as a stochastic variable, the three requirements for stochasticity being secured by the microdynamic rule and the sensitive dependence requirement being satisfied for the reason given by Maxwell, namely, that a molecule's state variables depend on its collisions, and the effects of its collisions depend on its precise position within a microsized region. (You can see why it is important, in this case, that the value of a stochastic variable need be independent only of the *approximate* values of the state variables: position is a state variable, and microposition is entirely determined by position—yet independent of approximate position.)

This is equidynamic business as usual: the microdynamic rule serves, I suggest, as the predominant source of stochastic variables for our applications of the equilibrium rule. That is, the stochastic variables we use are more often than not "low-level" versions of the state variables themselves, bearing the same relation to those variables as microposition bears to position.

What does that mean? As in the case of the bouncing balls, in the other kinds of systems to which the equilibrium rule is applied, the outcomes of the interactions on which the state variables depend are, because of their sensitivity to initial conditions, determined by the precise pre-interaction values of one or more state variables within a narrow range—precise position, precise direction of travel, precise speed, and so on. This precise-value-within-a-range, analogous to microposition, is the stochastic variable; its stochasticity is guaranteed by the microdynamic rule. Some examples, mostly invoking microposition itself, will be found in chapters 8, 9, and 11.

Let me provide some further exegesis of the equilibrium rule's requirements. First, if a variable that is in principle unbounded is for all practical purposes bounded, it may be considered to satisfy the equilibrium rule's requirement of boundedness. Consider molecular speed, for example: in a nonconservative system, there is no theoretical upper bound on speed, but some simple proportional dynamic reasoning suggests (correctly) that there is a maximum that is rarely, if ever, exceeded. Thus the equilibrium rule warrants the conclusion that the probability distribution over molecular speed will equilibrate.

Second, the equilibrium rule implies that all entities of a given kind have the same equilibrium distribution over the state variable in question. What determines sameness of kind? Two requirements must be satisfied for a set of entities to constitute a kind: the entities must have the same dynamics, meaning that their state variables should depend in the same way on the same stochastic variables, and the entities must occupy a *connected* space, meaning that a state that is accessible to one of the entities should be accessible to any of the others. Balls bouncing in a simple box occupy a connected space, for example, because any ball can reach any part of the box. If the box is divided by a wall, the space is not connected; the equilibrium rule must therefore be applied separately to the entities in the two halves, as though they were different kinds of things. This is clearly as it should be: the distribution of the positions of the balls in one half of the box will be different from the distribution of the positions of the balls in the other half.

Third, the rule as described contains some parameters with unspecified values. How sensitively must changes in the state variable depend on the stochastic variable? How close in space and time must values of a stochastic variable be to escape the demand for independence? And what period of relaxation time must be allowed before a system can be assumed to have equilibrated?

Good questions, all. The answers are connected: the more sensitive the dependence, and the closer in space and time values of the stochastic variable can be while not losing their independence, the shorter the relaxation time. (There are further determinants of the relaxation time, such as the size of the state variable space and others mentioned in section 7.4.)

The intuitive model we equidynamic reasoners have of an equilibrating system, or at least, of the sort of equilibrating system to which the equilibrium rule applies, is as follows. Each component of the system—each bouncing ball, say—is engaged in a random walk around its state variable space; that is, a journey through its allowed states (for a ball, the allowed positions and velocities) in which each next step depends on its present state and also, to a considerable extent, on a randomly determined quantity, namely, the present value of the stochastic variable on which the ball's trajectory sensitively depends. As the walk goes on, the ball's starting point, and more generally the system's starting point, makes less and less of a difference to the probability that the ball is in a particular state at a particular time. The influence of the initial conditions is "washed out," so that the probability distribution over the ball's velocity and position is eventually determined solely by the probability distribution over the stochastic variable and the form of the state's dependence on this variable. Whatever the starting point, then, the probability distribution over an entity's state variables finds an equilibrium determined by the fixed properties of the system (the shape of the box, for example), and the underlying microdynamics. So equidynamics disposes us to think.

We are rather good, I propose, at using this kind of model to reason about relaxation times and their relation to the randomizing power of a system's microdynamics, that is, their relation to the state variables' degree of sensitivity to the stochastic variable and the stochastic variable's independence properties—its intrinsic stochasticity, as it were. At the least, we can make educated guesses about relaxation time that are more often than not of the correct order of magnitude. This ability of ours is comparable to, indeed related to, our ability to perform proportional dynamic reasoning; as with that ability, I will take our know-how for granted rather than attempting to give it any further explanation (though see the end of section 7.5). Call this field of expertise *probabilistic dynamics,* over-general though that term might be.

The random walk model connects microdynamics to the notion of "equal opportunity" visitation rates described in section 6.3, and so to the notion of a shaking process; this idea is developed further in sections 8.1 and 8.2.

Why does the equilibrium rule work? The short answer is that our intuitive picture of the probabilistic dynamics of the systems falling under the rule is largely correct: the components of such systems are embarked on random walks that wash out the influence of the system's initial state.

Strevens (2003, section 4.4) fleshes out this answer by appealing to the ergodic theorem for Markov chains, the mathematicians' version of the "law of large numbers" for random walks, which requires for its application many of the same properties as the equilibrium rule: a microdynamics driven by conditionally independent probabilities, boundedness, connectedness, and so on. Further, the Markov justification provided by Strevens dovetails nicely with the concepts of microconstant and embedding processes introduced in sections 5.2 and 6.2; in deterministic systems where the justification applies, the underlying deterministic dynamics is an embedding dynamics, and thus the physical probabilities constituting the equilibrium distribution are founded in microconstancy.

To set up such a justification is, however, a complex task. For the application of the ergodic Markov theorem, the interaction dynamics must be represented as propelling the system on a random walk through a space in which each point represents a complete state of the system, as opposed to the state of a single component. The space must be coarse-grained, because microdynamic independence—constitutive of a Markov chain—holds only relative to such a coarse-graining, for reasons given in section 7.1. Finally, while the ergodic theorem guarantees convergence to an equilibrium distribution in the limit, a justification of the equilibrium rule requires something stronger, namely, convergence within the inferred relaxation period. Since I have given only the most general description of the way in which we determine relaxation periods, I am hardly in a position to provide this part of the story. For these reasons, I will not make any further attempt to explain the equilibrium rule's power.[4]

7.3 The Uniformity Rule

The equilibrium rule tells us that the probability distribution over a state variable will, with time, reach an equilibrium, but it gives us no clue as to the properties of the equilibrium distribution. Knowing merely that a global stable equilibrium distribution exists is of more practical use than you might think, but it would of course be more useful still to be able to infer the distribution's shape. That is the purpose of the uniformity rule, which spells out conditions under which the distribution may be supposed to be uniform, and in so doing supplies some clues as to how distributions over variables that fail to satisfy the conditions will deviate from uniformity.

The rule is as follows. Suppose that a system satisfies the conditions for the application of the equilibrium rule relative to some state variable x, so that x may be assumed to have an equilibrium distribution. Put a uniform probability distribution over every one of the system's state variables (not just x). If a state variable is unbounded, then put a uniform distribution over the range within which the parameter normally varies, and a zero distribution elsewhere. Determine how this distribution will change in the short term. If the distribution over x does not change—if it remains uniform—then infer that the equilibrium distribution over x is uniform, and that values of x for distinct entities, for example, the positions of different balls, are stochastically independent. If by contrast the distribution over x does change, departing from uniformity, then infer that the equilibrium distribution over x is not uniform.

Note that to forecast short-term changes in the imposed distribution, we must use our expertise in what I called in the previous section probabilistic dynamics—that is, our ability to model to some degree the dynamics of random walks in the systems to which the equilibrium rule applies.

You would apply the uniformity rule to a box of bouncing balls as follows. Put uniform distributions over the balls' position, direction of travel, and (reasonable values of) speed. Then do your best to see what will happen to those distributions as balls collide. As far as you will be able to tell, the distributions over position and direction will remain uniform: for every ball that is knocked out of one quadrant of the box into a neighboring quadrant, for example, another ball makes the reverse journey, on average. With speed, however, uniformity is not preserved: a ball that is traveling much faster than average is considerably more likely to experience a decelerating interaction than an accelerating interaction, while the contrary is true for a ball traveling much more slowly than average. You conclude that in the long term, the distribution over position and direction of travel will equilibrate to uniformity, while the distribution over speed will not. (Also nonuniform, according to the test, are various nonstandard quantifications of position and direction of travel.)

There is a corollary to the uniformity rule, I propose, that allows its user to extrapolate from the way in which a state variable fails the uniformity test to conclusions about the way in which that variable's equilibrium distribution (guaranteed by the equilibrium rule) deviates from uniformity: the equilibrium distribution will typically differ from the uniform distribution in the same way that the putative distribution imposed in the uniformity test tends to diverge in the short term from uniformity.

For example, since a uniform distribution putatively imposed over the speed of bouncing balls deviates from uniformity in the short term by the thinning of its tails—very fast balls tend to get slower and very slow balls tend to get faster—you can conclude (defeasibly as always) that the equilibrium distribution over speed will be fatter in the middle than at the ends.

Or suppose that the balls are subject to a short-range force pulling all balls toward the bottom of the box, but acting more strongly on balls near the bottom than on balls near the top. A putative uniform distribution over position will soon deviate from uniformity: in the short term, the top of the box will become more sparsely populated, while the bottom's population increases. You can infer that the equilibrium distribution over position will have the same property, so that in the long term, a ball is more likely to be found at the bottom than at the top. (The same reasoning applies, of course, to a gas contained by a gravitational field, such as a planetary atmosphere.)

On to the justification of the uniformity rule: why does it work? Consider a system with only one state variable. Suppose that the equilibrium rule applies, and correctly determines that the variable has a probabilistic equilibrium—that in the long term, the state of every entity in the system falls under a single probability distribution.

Suppose also that the variable passes the uniformity test: a uniform distribution over the state variable remains uniform in the short term. Since the long term is nothing more than a succession of short terms, it follows that the uniform distribution is an equilibrium. But the equilibrium rule implies that there can be only one equilibrium, the globally stable equilibrium whose existence is asserted by the rule. Thus, the globally stable equilibrium is the uniform distribution.

Now suppose that the variable fails the uniformity test. Then the uniform distribution cannot be an equilibrium distribution, thus is not the globally stable equilibrium promulgated by the equilibrium rule. (You will recognize in this argument a line of thought I attributed to Maxwell in section 2.3.)

When there are two or more state variables, the justification is not so straightforward. Suppose that a system has two state variables x and y, and that at equilibrium, x has a uniform distribution and y has a certain nonuniform distribution. Suppose further that the stability of x's uniform distribution depends on y's distribution being close to its equilibrium, or at any rate being something other than uniform. Then in the uniformity test, when a putative uniform distribution is put over both variables, x's distribution will deviate from uniformity. The test will therefore deliver the incorrect judgment that x's equilibrium distribution is nonuniform. The same sort of problem may arise in cases where the distribution of y does not equilibrate at all. I am unsure whether our equi-dynamic reasoning is vulnerable to false negatives of this sort, or whether there is some further component to the uniformity rule that I have omitted.

Also in need of justification are, first, the independence component of the uniformity rule, which may follow immediately from a similar com-ponent in the equilibrium rule (about which I earlier declared agnosti-cism), and second, the corollary to the rule, which seems to rest on the assumption that a probability distribution's approach to equilibrium is simple and deliberate, so that early moves toward equilibrium from a particular state (uniformity) are indicative of the overall equilibrating trend. But let me put these further questions aside in order to move on to issues concerning the scope and historical importance of equidynamic reasoning.

7.4 Bouncing Further Afield

How do we deal with shapes other than spheres? In Téglás et al.'s (2007) experiments with infants, the visual stimuli are objects with rather com-plicated geometries. You might also wonder about "molecular" shapes, that is, shapes that have the contours of diatomic gas molecules such as oxygen and nitrogen, or more complex molecules such as carbon dioxide, methane, or ammonia (figure 7.1).

The key questions here are, first, whether the microdynamic rule ap-plies to collisions between objects with these geometries, and second, whether the uniformity rule applies to their position and direction of travel. The answers in both cases are affirmative. The microdynamic rule applies because the collisions will be sensitive to the pre-collision state and because the colliders' post-collision state—their position, transla-tional velocity, and rotational velocity—will depend smoothly if not uniformly on their pre-collision state. The uniformity rule applies be-cause a uniform distribution over the orientation and so on of colliding molecules will remain stable in the short term, since collisions so distrib-uted do not favor any particular positions or directions of travel. Note that with complex bouncers, there will be no Maxwellian equiprobabil-ity of rebound angle (see section 2.3); for the application of the rules formulated in this chapter, however, none is required.

There are many other sophisticated aspects of our thinking about bouncing scenarios that might be explained along the lines laid out

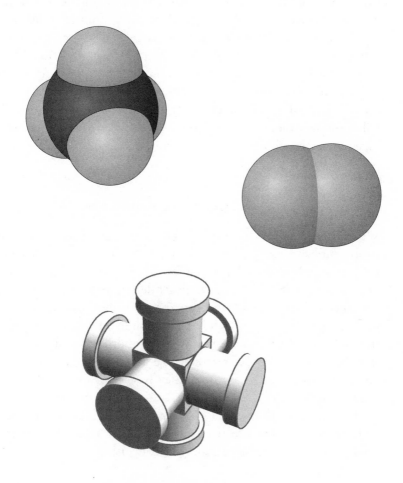

Figure 7.1: Some bouncers with complex, nonspherical geometries.
From top: methane; diatomic nitrogen; a shape used by Téglás et al. (2007).

above. A good proportion of these are connected to our use of probabilistic dynamics to estimate relaxation time. For example, as with the urn, we seem to have at least a rough quantitative sense of how far a ball can be expected to travel in a given time, seeing for example that travel will be slower in denser gases. Another related and revealing case is a box bisected by a wall with a hole through which individual balls only barely fit. Suppose that a ball starts on one side; how much time must elapse before it is equally likely to be found on either side? We can make a well-educated guess, and certainly, we can see that the probability distribution over the ball's position will eventually equalize, though it will take far longer to do so than if the hole were larger or the wall were altogether absent.

7.5 Children on Bouncing

What about the children? Are Téglás et al.'s three-year-olds, are their twelve-month-year-olds, proficient in the use of the microdynamic rule, the equilibrium rule, the uniformity rule? Or do they estimate the bouncing probabilities using some gentler, simpler rules of reasoning?

Earlier I distinguished two broad approaches, dynamic and nondynamic, to understanding children's reasoning about urn-drawings (section 6.4). According to the nondynamic approach, children do not derive the movement of balls around the urn from the physical principles governing ball dynamics; rather, they base their inferences about ball-drawing probabilities on the urn's static properties, most notably, the haphazard (or in a loose and popular sense, "random") arrangement of balls in the urn. In this section, I ask whether a related nondynamic hypothesis can make sense of children's conclusions about bouncing balls' exit probabilities.

The nondynamic hypothesis I have in mind for the case of bouncing hinges on the child's observation of the haphazard motion of the ball or balls in the box. (It is nondynamic because the children do not make use of any knowledge about the physical foundation of the haphazardness.) The child takes this pattern of motion, according to the hypothesis, as grounds for inferring that events depending on the motion should have a Bernoulli distribution, with probabilities proportional to the possibilities. For example, when watching a ball bounce around a box with three exits on the right side and a single exit on the left (figure 4.1), they will expect a right-side exit to be three times as likely as a left-side exit.[5]

Presumably, children do not have a fixed inventory of possible outcomes—of container shapes and exits, for example—for which the nondynamic rule recommends equal probabilities. Rather, the rule must provide some sort of recipe for determining when ascriptions of equiprobability are merited, that is, some sort of criterion according to which, for example, the apertures in the multiple-exit box are all sufficiently similar that, given haphazard motion, exits from each are equally probable.

That criterion, if it is at all general, will surely invoke proportional dynamics: children will base their judgment of relevant similarity on the fact that the positions and directions of travel that will take the ball out one exit occupy a similar proportion of the possibilities as the positions and directions of travel that will take it out any other. But then why not suppose that children apply the same sort of reasoning elsewhere, not only evaluating the probabilities of various final outcomes but also of various interactions that might occur on the way to those outcomes? If, for example, they can use the rule to infer the probability that the ball will exit through a certain aperture, they can presumably use it also to infer the probability that, after a certain period of time, the ball is in the left-hand half of the box, or the probability that, after a certain period of time, the horizontal component of the ball's motion is leftward rather than rightward—saying, I suppose, that the probability is in each case one-half. A sufficiently general and flexible "nondynamic" rule is easily manipulated, it seems, to pronounce on many aspects of the dynamics of bouncing. In that case, the nondynamic hypothesis is rather dynamic after all—and what it attributes to children is perhaps not so different from the microdynamic and equilibrium rules.

Are there dynamic rules other than the microdynamic and equilibrium rules, then, that children might use? Téglás et al. (2011) propose a rule in which children infer probabilities for various events in the bouncing scenario by running simulations of several episodes of bouncing and inferring the probability of an event from its frequency in the simulated episodes (a kind of Monte Carlo procedure). How, you might wonder, could this rule be plausibly extended to the case of the urn, where the dynamics are far more complex than in the case of the four objects bouncing in a box? Even in the case of the box, the authors use a simplified dynamics in which they assume that the objects move by Brownian motion, rather than explicitly calculating the effects of collisions. Perhaps the same assumption could be used for urns and other such setups?

If so, a further rule is needed to determine to which systems it is reasonable to apply a Brownian model. But this is, of course, precisely the sort of thing that the microdynamic and equilibrium rules are good for. A synthesis therefore suggests itself: my microdynamic and equilibrium rules are used by child and adult reasoners to determine that a system can be simulated by way of a certain kind of stochastic model, and Téglás et al.'s procedure is then applied, using that model, to implement what I have called probabilistic dynamic reasoning.

8

UNIFYING

Humans' equidynamic inferences about bouncing balls were explained in chapter 7 as guided by three rules: the microdynamic rule, the equilibrium rule, and the uniformity rule. Considered as a unified inferential strategy, call this trio the *equilibrium rule package*. The package is used in this chapter to explain other aspects of equidynamic inference, taking care of ends left loose in the earlier discussions of tumbling dice and urn drawings, asking some questions about the complexity of equidynamic inference, and finally making sense of Maxwell's spectacular success.

8.1 Dice Revisited

The uniform distribution over the outcomes of the canonical die roll described in section 6.2 is due, I argued, to the die's being shaken before its release. This motion is the paradigm of a shaking process, that is, a process in which an outcome-determining "shaking parameter" wanders through its space of possible values in a way that is irregular yet probabilistically egalitarian, in the sense that, once a certain time has elapsed, it is equally likely to have taken on any of those values. Microconstancy supplies a physical basis for the uniform probability distribution over "visitation rate," but the reasoning process by which we get our equidynamic grip on visitation rates, I suggested in section 6.3, invokes not the property of microconstancy per se but rather the idea that the die, in the course of its crooked path back and forth through the space between my hands, executes a random walk.

There I left the story; let me now use the equilibrium rule package to supply the missing parts. Suppose you have chosen as your shaking parameter the die's axis of rotation. Your aim, then, is to infer a probability distribution over the axis around which the die, as it flies from your hands, is spinning. (From a uniform distribution over rotational axes you will then use proportional dynamics, as in the case of the bouncing coin [section 6.1], to infer a uniform distribution over the roll's six possible ultimate outcomes.)

You reason, I propose, as follows: the die's rotation axis on exit is largely determined by a series of collisions with its container walls—your hands. Noting the die's relatively smooth collision dynamics and the sensitivity of its post-collision state to its pre-collision state, you use the microdynamic rule to infer that position and direction of travel are microequiprobabilistically distributed in the short term.

You then invoke the equilibrium rule, using as stochastic variables either microposition or microdirection (the "micro" version of direction of travel; that is, the precise direction of travel within a narrow band of possibilities), followed by the uniformity rule, from which you infer a uniform equilibrium distribution in the long run over those variables that pass the rule's stability test: position, direction of travel, and axis of rotation, but not speed, either translational or rotational. (I leave it to you to verify that the rules apply to the case of the die as claimed; the reasons are much the same as with the bouncing balls.) You thereby conclude that the die is equally likely to have any axis of rotation at any time after a certain relaxation period has elapsed. If the shaking goes on for long enough, then, there is a uniform probability distribution over the die's axis of rotation at the moment of release.

Along with this conclusion comes a model of the shaking process according to which the die's rotation undergoes a random change determined by its microposition, microdirection, or both, each time it hits its container walls, so that after relaxation, it is equally likely to have any value, or in other words, so that after relaxation, it is equally likely to visit each possible axis—what I earlier called equal opportunity visitation. This probabilistic dynamical model is not merely a theoretical idyll; it is what allows you, by way of a back-of-the-mental-envelope calculation, to estimate the relevant relaxation period, that is, the degree of shaking that is necessary to equiprobabilify the rotation axis by the time of the die's release.

The relaxation period commences, you infer, once the shaking has created a sufficient degree of randomization. In the case of the die, you might compute a rough degree of randomization by multiplying the initial condition sensitivity of a single die-wall collision by the number of such collisions preceding release. If this number is high enough, you have relaxation and therefore a truly randomizing shaking; otherwise, the randomness of the roll as a whole will depend on how roughly or carefully the die is set (oriented before shaking), and how wildly it tumbles when it hits the table.

In this way, the hypothesis that we reason using the equilibrium rule package explains why we infer a uniform probability distribution over the exiting die's axis of rotation but not, say, over its velocity or over some Bertrandian function of axis of rotation that favors axes that pierce the 1 and 6 sides of the die (since neither velocity nor the Bertrandian variable will pass the uniformity test).

A further way to test the explanatory power of the equilibrium rule package is to ask under what circumstances it predicts that we will refrain from inferring a uniform distribution over rotation axis. Further revealing questions: When, though we refrain from inferring a uniform distribution, do we concur that there is some fixed long-term distribution over a parameter? What qualitative properties of that distribution can we infer?

Two cases are of more practical importance than the rest. In the first, the die is shaken only briefly, perhaps colliding with its container walls just once before it is released. As noted above, the equilibrium rule package predicts our reluctance to count such a roll as imposing a uniform distribution over the possible outcomes, unless some other facet of the process—such as a bouncing vigorous enough to qualify as a shaking in its own right—makes up the deficit of randomization.

In the second case, the die is loaded. In spite of the loading, the microdynamic rule can be applied: you can infer that even a loaded die has a microequiprobabilistically distributed position. Consequently, you can apply the equilibrium rule to infer that, once a relaxation period of suitable length has passed, there is a long-term probability distribution over the die's axis of rotation. Is the distribution uniform? Apply the uniformity rule. Since your physical intuition, or more exactly your skill with probabilistic dynamics, tells you that a uniform distribution over a loaded die's axis of rotation is not—or at least, may not be—stable in the short term, you can infer that the distribution is not, or may not be, uniform in the long term. Shake a loaded die as long as you like, then; you will not equiprobabilify its axis of orientation.

Can you infer, equidynamically, anything about the outcomes of a toss of the loaded die? You know that there is a stable physical probability distribution of some sort over the die's properties as it leaves your hands. The later stages of the rolling process—the die's flight through the air and its bouncing on the table—will transform this into a distribution over the outcomes of the roll.[1] You can therefore conclude that there is a fixed physical probability distribution over the outcomes of a loaded die roll. This is useful information. In particular, it gives you confidence that past frequencies of outcomes will very likely predict future frequencies or, more revealingly, that the best hypotheses about loaded die rolls are those that ascribe a Bernoulli distribution to the outcomes, a conclusion that greatly simplifies your hypothesis space and thus your testing procedure: you know that all you need observe when experimenting with the die are the frequencies with which the different outcomes occur.[2]

If your physical intuition, and in particular your command of proportional dynamics, is sufficiently sophisticated, you might be able to go further. You might infer that the distribution over the die's axis of rotation after shaking will favor those axes that do not pierce the loaded side of the die, and you might even infer that once the die is tumbling on the table, it is more likely to land with its loaded face downward, allowing a qualitative conclusion about which faces are more probable, which less probable, in a loaded roll.

8.2 Urns Revisited

Perhaps the most striking, or alarming, feature of the urn is how complex a system it is by comparison with the shaken die: when shaking an urn full of balls, perhaps 100 objects are in a constant state of n-way interaction, for n often much greater than 2. Yet the urn is as transparent to our physical probabilistic intuition as the die roll: it is as effortlessly clear to us that a well-shaken urn will yield any ball with equal probability, as it is that a well-shaken die will with equal probability yield any face.

An account of equidynamic reasoning should make sense of this psychological datum; in particular, it should attribute to us a form of reasoning whose complexity does not increase as the number of degrees of freedom in the system in question increases, at least when the moving parts are relevantly similar. The hypothesis that we reason using the equilibrium rule package satisfies this desideratum, giving the following account of our reasoning about the sort of urn-drawing described in section 6.4, in which it is always the ball closest to a fixed "target zone" that is drawn.

First, you apply the microdynamic rule to determine that the probability distribution over any ball's position and direction of travel during shaking is microequiprobabilistic. This gives you, in the form of microposition or microdirection, the stochastic variables you need to apply the equilibrium rule, and so to apply the uniformity rule, deriving a uniform distribution over position, thus an equal probability of being drawn, for any ball—if the shaking goes on for long enough. (Again, I omit the details, which are not deeply different from the case of the bouncing balls.) The number of balls involved makes no difference to the complexity of the inference.

What if the shaking period is too short to give every ball an equal chance to enter the target zone? You then use probabilistic dynamics, taking into account such parameters as the vigor of the shaking, to distinguish between a case in which the original arrangement of balls is haphazard enough for a short shaking to randomize the relevant probabilities, so that the probability of a color being chosen is proportional to the number of balls of that color, and a case in which, due to a highly ordered initial arrangement, color will not be fully randomized.

The reasoning involved in making such a judgment estimates the distance that a ball will typically travel while the urn is shaken, and so the potential of the shaking to send a representative sample of balls through the target zone, by assuming that shaking propels each ball on a random walk around the urn. This is the same kind of probabilistic dynamic reasoning that a user of the equilibrium rule package deploys to estimate the relaxation period when inferring the existence (or otherwise) of an equilibrium distribution over the properties of chapter 7's bouncing balls. The equilibrium rule package is therefore easily able to explain our facility with short shakings; no new inference procedures need be postulated.

Consider next a complication comparable to the loading of a die. Suppose that a drawing from an urn is made with a magnetic claw, and that the urn contains half red balls with an iron core and half white balls without the core (though otherwise of the same mass and so on). Intuitively, a red ball is more likely to be drawn than a white ball. What reasoning goes into this conclusion?

Because the red and white balls are physically identical apart from their responsiveness to the magnet, we can infer that they pass through the target zone with equal probability. The bias in favor of red is entirely attributable to the process that plucks a ball from the zone. To determine the direction of this bias, we presumably reason that more configurations of white and red balls in the target zone will lead to a red ball's being chosen than a white ball's being chosen. We think explicitly, then, about what proportion of initial conditions lead to one kind of outcome and what proportion to another—that is, we engage in what I have called proportional dynamic reasoning.

Proportional dynamic reasoning may also play a role, I suggested, in our thinking about the tumbling of a loaded die (section 8.1); the same is surely true of our equidynamic wrangling with asymmetries in various other probabilistic processes, such as the wheel of fortune with little braking pads under its red sectors mentioned in section 5.1. At the end of section 6.1, I wondered whether we reason explicitly about the effect of a tossed coin's rough landing, determining that the bounce will make no difference to the proportion of initial conditions leading to heads rather than tails. Our facility with proportional dynamic reasoning in the case of the urn and the die suggests an affirmative answer.

8.3 Complexity

The equilibrium rule package has several moving parts, each of which itself has further parts. Is all this complexity necessary, you might wonder, to explain our equidynamic reasoning? Various convolutions have been justified at various points along the way, but it may be worth pausing to face this niggling question squarely, gathering together in one place some of the more important excuses for the aggregate level of intricacy so far attained.

To that end, consider several ways in which the equilibrium rule package complicates itself. First, the package provides rules for inferring both short-term and long-term probability distributions over the parameters of the system to which it is applied: a microequiprobabilistic distribution in the short term, and some sort of equilibrium distribution (perhaps uniform) in the long term. Why not omit the short-term distribution, preferring a rule that goes from physical properties directly to long-term distributions? After all, the physical premises needed for the short-term and long-term inferences are rather similar; could they not be merged?

No; our equidynamic reasoning manifestly makes use of the short-term distribution (or something similar and no simpler). Most obviously, reasoning about relaxation periods—without which the long-term equilibrium rule is uninformative—is based on a grasp of pre-equilibration dynamics, hence on some kind of short-term dynamics, which grasp must surely be probabilistic. When, for example, we ask whether a box full of balls has been sufficiently shaken to randomize its initially organized state, we imagine (I have proposed) each of the balls in the box taking a random walk, and we ask ourselves whether the walk goes on for long enough to scramble the balls. To understand the walk as random—as making stochastically induced twists and turns as it wends its winding way through the space of possible states—requires the postulation of a short-term probability distribution over ball dynamics.

This raises the question whether the equilibrium rule asks for too much in requiring that we think explicitly about the dynamics of the processes to which it is applied, using both proportional and probabilistic dynamics. Is this blending of measure and motion worth the mental effort? Could we not base an effective system of structure-to-probability reasoning on physical structure without taking the trouble to compute the dynamic consequences of that structure?

There might be a form of structure-to-probability reasoning that ignores dynamics, but it would be far less sophisticated than our own. Observe, for example, that you reason explicitly about dynamics when you infer that an "army blanket roll"—a die roll in which the die rotates around a more or less fixed axis perpendicular to two of its faces—will not yield each of the six possible outcomes with equal probabilities. The same is true of a ball drawn from an urn by a magnetic claw, as described in the previous section. Even young children's inferences about urn and bouncing probabilities, I pointed out in sections 6.4 and 7.5 respectively, appear to have a dynamic component.

Further, it is quite unclear, if dynamic considerations are ignored, what should count as a relevant structural property, or in particular, as a relevant symmetry—which brings me to the next and last complaint.

The final complication in the equilibrium rule package is that its tests for relevant physical structure—not least, the microdynamic rule's smoothness condition and the uniformity rule's stability test—are themselves moderately elaborate. Is there not something simpler?

The question which structural properties are relevant—that is, which properties license inferences to microequiprobability, equiprobability, and so on—is more difficult than it looks. A spherical ball is bouncing around a box. But the box itself is asymmetrical. Does it matter? Does the irregular shape of our cupped hands matter when we shake a die? And what about the ball itself? None of the "bouncers" in figure 7.1 has the symmetry of a sphere; what probabilistic inferences, if any, can we base upon their shapes? Any rule capable of answering these questions intelligently must be at least some what complex.

8.4 Maxwell Reconstructed

How did Maxwell succeed in inferring the correct probability distribution over the velocity of gas molecules without velocity statistics? And why did his derivation of the distribution have the persuasive power that it did—not irresistible, perhaps, but sufficiently convincing to bear the weight of, and in particular to recruit powerful scientific minds to, an entirely new and ultimately enormously successful research program in physics?

These questions are complicated by the existence of two discernible trains of thought in Maxwell's 1859 paper, what I have called the official and the unofficial derivations of the velocity distribution. The official derivation, contained entirely in Maxwell's proposition IV, attempts to establish two properties of the distribution—the equiprobability of a velocity's direction given its magnitude, and the independence of a velocity's Cartesian components—on a priori grounds. A mathematical argument (possibly borrowed from Herschel) then shows that only a single distribution satisfies these joint demands.

The unofficial derivation is revealed only with the help of some textual detective work. In section 2.3, I offered the following partial reconstruction.

From the assumed microequiprobability of position in the short term, Maxwell derives a uniform short-term distribution over post-collision rebound angle (proposition II). He then reasons, in a passage that is somehow both persuasive and obscure, that the distribution of the velocities and positions of gas molecules will tend toward a single, stable equilibrium (proposition III).

Proposition IV opens with the declared intent to deduce the form of the equilibrium velocity distribution. The premises of the official, Herschelian argument for the velocity distribution are at this point stated and given their aprioristic justification in terms of conceptual truths about Cartesian coordinates. But—and here is the most speculative element of my reconstruction—lurking in the back of Maxwell's mind and text is an alternative justification for the Herschelian posits, based on the fruits of the reasoning in propositions I through III: the probabilistic dynamics of collisions and the existence of a global, stable equilibrium distribution. In that alternative derivation, which is perhaps Maxwell's original derivation, it is first noted that any bias among molecules with a given speed in favor of some directions of travel over others, or correlation between the Cartesian components of velocity, will tend to change in the short term, and it is then observed that a distribution that changes in the short term cannot be a stable equilibrium, as explained in section 2.3. It follows that in the equilibrium velocity distribution, the Herschelian posits hold.

Maxwell's unofficial derivation of the velocity distribution for the most part runs parallel, I propose, to the natural course of human equidynamic reasoning, as schematized by the equilibrium rule package. Further, Maxwell vaulted the logical chasms in his unofficial argument thanks to the propulsive cognitive power of the equidynamic rules. At the most critical points in his derivation he was, in other words, falling back on instinctive human inferential tendencies—he was thinking like a twelve-month-old, if rather deeper and rather harder. Let me show you how.

Maxwell's first statistical premise about molecules, that their position is microequiprobabilistically distributed, of course comes straight from equidynamics, by way of the microdynamic rule.

The argument for equilibration is based on the equilibrium rule, with rebound angle playing the role of the stochastic variable. (Maxwell's derivation foregrounds the equiprobability of rebound angle, but as noted in section 7.4, the equilibrium rule does not require uniformity of the stochastic variable's distribution—and just as well, since rebound angle equiprobability does not hold for nonspherical molecules.) This explains why an argument that is from a logical or mathematical point of view clearly incomplete nevertheless has considerable persuasive power for Maxwell, for his readers, and for us.

What of the interpolated elements of the unofficial argument, which take Maxwell (on my interpretation) from the probabilistic collision dynamics and the fact of global equilibration to proposition IV's two Herschelian posits?

They are not, I think, a proper part of the equidynamic method. What equidynamics provides directly is, by way of the uniformity rule, the equiprobability of position and direction of travel—useful and suggestive, but not sufficient for the derivation of the Herschelian posits.

Equidynamics also gives us, however, expertise in what I have called probabilistic dynamics, an important part of which is the ability, used in the application of the uniformity rule, to predict the short-term direction of random walks. This cognitive facility plays an essential part in the unofficial argument; it is used to infer a dissociation of any correlation between, first, the magnitude and direction of, and second, the Cartesian components of, molecular velocity.[3] The tendency to Cartesian dissociation is, as I remarked in section 2.3, harder to perceive than the dissociation of direction and speed, but it is nevertheless both real and within the epistemic reach of a talented equidynamic reasoner.

What is needed to complete the justification of the Herschelian posits is the observation that short-term trends contain information about the long-term equilibrium; here, Maxwell was on his own. (I used the connection between the short term and the long term to justify the uniformity rule in section 7.3, but the justification is not a part of equidynamics per se.)

This is the explanation of Maxwell's success, then. His cognitive birthright as *Homo sapiens* comprised among many other things a form of physical intuition implemented by the suite of inferential rules that constitutes equidynamics, most particularly the microdynamic rule and the equilibrium rule, which gave him the probabilistic premises that served as his starting point—namely, microequiprobability of position and the existence of a fixed long-term velocity distribution. Some further keen insights about the probabilistic dynamics of molecular collisions, themselves applications or adaptations of equidynamic thinking, secured the Herschelian posits. The rest was simple mathematics. Equidynamic thinking is not foolproof, but it is quite reliable, and in this case as in many others, it delivered the probabilistic goods.

Why was Maxwell's argument as persuasive as it was—and no more? Maxwell's readers may be divided into two groups: those who appreciated the force of the unofficial derivation, whether thanks to a close reading of the text or their own reconstruction of the equidynamics, and those who saw only the official derivation. The argument must surely have failed to impress the latter group, though they might have been willing to play along with Maxwell in the hypothetico-deductive spirit, on the grounds that his conclusions at least resemble the conclusions fully warranted by equidynamics, such as the equiprobability of direction of travel.

Readers who perceived the equidynamic structure of Maxwell's unofficial derivation would have had to assess the plausibility of the following elements of the argument:

1. Molecular position is distributed microequiprobabilistically,
2. Molecular velocities converge to a unique equilibrium distribution,
3. There is a short-term tendency for direction and magnitude of velocity to dissociate,
4. There is a short-term tendency for Cartesian components of velocity to dissociate, and
5. The short-term tendencies and the existence of a global equilibrium distribution imply corresponding long-term independences.

As humans, and therefore as equidynamic reasoners, the reader would find elements (1) and (2) eminently plausible—as do we, even if we cannot explain why. To grasp (5) requires an aptitude for probabilistic reasoning. That leaves (3) and (4). Both have a certain plausibility. But (3) is, as I remarked above, not only easier to perceive, but is made apparent by Maxwell's reasoning in propositions II and III in a way that (4) is not.

Consequently, you should expect that elements (1) and (2) will be passed over with little comment, as though they are unremarkable, but that the Herschelian posits based on (3) and (4) will attract more scrutiny. Of these, the latter—the independence of the Cartesian components of velocity—will seem the more tendentious.

Maxwell's readers, then, along with Maxwell himself, had at least moderately good reasons to think that Maxwell's velocity distribution was correct. Any doubts they had would most likely be directed at the assumption of component independence. That, you may recall, was Maxwell's own later reaction: he singled out the assumption as "precarious" (section 2.1).

The mystery posed in chapter 1 is therefore solved. Maxwell's derivation of the velocity distribution was not a priori: its reasoning took as its premises empirical facts, though so quietly that their contribution has been overlooked, not only by historians but perhaps also by the original deriver himself. How could that be? Maxwell was uncommonly brilliant, but he was to some extent a spectator to his own success. The probabilistic genius responsible for his velocity distribution is not a person but a habit of thought shared by every one of us, the equidynamic method. This method, with its additional and unofficial supplement—for which Maxwell alone deserves the credit, even though he might have dwelled on it longer and made it more explicit—both uncovered the truth about the velocity distribution and convinced sufficiently many of Maxwell's readers of its merits that his discovery was not overlooked or buried but rather played a central role in the creation of statistical physics.

Why stop there? I observed in section 2.3 that the structure of the unofficial argument for proposition IV's Herschelian posits appears explicitly in Maxwell's argument, later in the same paper, for the equipartition of energy, that is, for the tendency of kinetic energy to equilibrate equally among all degrees of freedom (in propositions VI and XXIII).[4] Maxwell, Boltzmann, and Gibbs later made equipartition the foundation of all of statistical mechanics.

Could equidynamics be the philosopher's stone of statistical physics—transmuting, at every critical point in its development, structural premises into probabilistic conclusions? That might explain why J. J. Waterston in 1851 so confidently stated the equipartition hypothesis for the first time in human history, in an otherwise entirely nonstatistical context (Truesdell 1975). It might explain the plausibility of the brute statistical posits of Maxwell's (1867) attempt to rebuild his kinetic theory. It might explain Boltzmann's (1964) reliance on his otherwise foundationless *Stosszahlansatz,* or assumption of molecular chaos—which assumption is a consequence of the microequiprobability of position.

The entire history of statistical physics might be told, in other words, as a series of seemingly impossible intellectual leaps made feasible by the invisible wire of equidynamics running from prehistoric times to the present.

III

BEYOND PHYSICS

9

1859 AGAIN

Statistical physics was not 1859's only theoretical innovation. *On the Origin of Species* set alight a star in biology to outshine even Maxwell's achievement in the physical sciences. As well as its birthday, evolutionary theory shares with statistical physics a deep and essential foundation in equidynamic thinking.[1]

This is not obvious, you are thinking. Maxwell's 1859 paper begins with the postulation of a probability distribution. Darwin's 1859 book contains no probability distributions, indeed, virtually no quantitative thinking at all. But not all probabilistic thinking is quantitative thinking; some is qualitative or comparative. In this sense, almost all of evolutionary biology is run through with probability, if not transparently so. The explicitly stochastic aspects of modern evolutionary theory—most notably the mathematical theory of drift—are not statistical islands but mountain peaks, revealing the presence of continents of probabilistic cogitation under the clouds.

9.1 The Stochastic Nature of Selection

Three conditions jointly satisfied bring about a population's evolution by natural selection, it is traditionally said. First, the population should consist of individuals who reproduce in such a way that the offspring tend to resemble the parents. Second, the individuals in the population should vary in their physical makeup. Third, these differences in makeup should bring about differences in the individuals' life spans or rates of reproduction; those with longer life spans or rates of reproduction are then said to be fitter than the rest. Under these circumstances,

127

the individuals with greater fitness will tend to take over the population; the population will become dominated, then, by the fittest available traits.[2]

Probabilistic processes contribute to or otherwise buffet the course of evolution by natural selection under all three headings above. They affect inheritance because, among other reasons, the transmission of genetic material in sexual reproduction is inherently stochastic: the genes that a parent contributes to a particular offspring are, loosely speaking, a random sample of the total parental genotype. They affect variation because, among other reasons, mutations of a genotype are caused by inherently stochastic processes, such as chance strikes by cosmic rays and other environmental radiation. And they affect survival and reproduction rates because in the chaos of a typical habitat, almost every strategy for biological advancement is shot through with risk: the risk of predation, the risk of disease, the risk of starvation, and the many risks of various mating gambits.

It is the last of these manifestations of stochasticity in natural selection that will provide the foundation for this chapter. Almost any trait that is evolutionarily advantageous—that extends the time during which an organism may produce or nurture its offspring, or the rate or quality of such production or nurturance—varies in its effectiveness. Fitness is *average* advantage, or *expected* advantage; to compare fitnesses, and so to predict or explain one trait's coming to dominate and eventually to drive to extinction another trait, is therefore to compare average contributions to survival and reproduction.

Comparisons of averages need not make use of explicitly stochastic representations, but such representations are a natural tool for thinking about fitness. And they have been used in just this way, in the earliest expositions of evolution by natural selection and, in probabilistic models of selection and drift in population genetics and elsewhere, in modern evolutionary theory as well. It is the aim of the next two chapters to show how stochastic thinking about fitness, both historical and modern, relies on equidynamic reasoning to compare the fitnesses of different traits, to infer the likely form of explanatory models for particular episodes of evolutionary change, and to generate new hypotheses about the causes of such change.

9.2 Swift Wolves

"In order to make it clear how . . . natural selection acts," Darwin develops two hypothetical examples in chapter 4 of the *Origin* (Darwin 1859, 90–96). The first of these "imaginary illustrations":[3]

> Let us take the case of a wolf, which preys on various animals, securing some by craft, some by strength, and some by fleetness; and let us suppose that the fleetest prey, a deer for instance, had from any change in the country increased in numbers, or that other prey had decreased in numbers, during that season of the year when the wolf is hardest pressed for food. I can under such circumstances see no reason to doubt that the swiftest and slimmest wolves would have the best chance of surviving, and so be preserved or selected.

Darwin recognizes that, because of the vicissitudes of the lupine life, even under these favorable circumstances the swiftest and slimmest wolves may not do better than the rest. Their advantage is probabilistic: they have the "best chance" of surviving. Elsewhere Darwin writes in the same vein, typically using the word *chance*.[4] In the first paragraph of the chapter on natural selection, for example, he characterizes the process as follows:

> Can we doubt . . . that individuals having any advantage, however slight, over others, would have the best chance of surviving and of procreating their kind? (80–81)

Further, in the sixth edition of the *Origin,* Darwin inserted an additional paragraph into chapter 4 commenting on the possibility that, by sheer bad luck, a fitter variant might fail to prosper.[5]

Despite the haphazard course of any individual wolf's life, Darwin does not hesitate to predict the fate of wolves in general: from the swiftest and slimmest wolves' having the best chance of surviving, he infers that they will increase in numbers relative to their peers—they will be selected. To all appearances, then, Darwin is appealing to the law of large numbers. He is attributing to different varieties of wolf different probabilities of surviving and reproducing in the conditions specified in his scenario, and he is predicting that, because frequencies tend to follow probabilities, the variety with the higher survival and reproduction probabilities will grow faster (or shrink more slowly) than the rest.

Unlike Maxwell, Darwin was no mathematician, and as I have already remarked, he does not explicitly call on the law of large numbers or any other apparatus of probability theory in the *Origin*.[6] But he had the same informal knowledge of the long-term effect of causal biases in stochastic processes, such as the loading of a die or the placement of braking pads under the red sections of a wheel of fortune, as any human being, not least one who had spent a "good deal" of his Cambridge evenings at the blackjack table (Browne 1995, 98). It is this knowledge that he is putting to work in reasoning about his wolves.

The same stochastic interpretation of Darwin's thinking about natural selection is urged by Sheynin (1980, section 5.4) and Hodge (1987, 246).[7] Few historians have disagreed.[8] I take it as the basis for everything that is to follow.

Assume, then, that like modern evolutionary biologists, Darwin conceived of the dynamics of the wolf population in stochastic terms. His inference about the wolves in his hypothetical scenario might then be reconstructed roughly as follows:

1. A wolf's probability of capturing the newly predominant prey, deer, increases as it becomes swifter, therefore
2. A wolf's expected life span increases as it becomes swifter (as perhaps does its ability to nurture its young), therefore
3. A wolf's expected number of offspring increases as it becomes swifter, therefore
4. Typically (here the law of large numbers is implicitly applied), swifter wolves will have more offspring.

Such an inference has the same structure, I insinuated above, as an inference in which you conclude that a wheel of fortune with braking pads under the red sections will yield *red* more often than one without, or that a loaded die will yield some outcomes more often and some less often than a fair die. With such inferences, it shares four notable features.

First, it is characteristically silent. Just as an ordinary, mathematically unsophisticated adult will have difficulty finding the concepts to explain their inferences about loaded dice or biased wheels of fortune, while insisting on the conclusions for the right reasons all the same, Darwin does not spell out his reasoning about the wolves and other cases of natural selection. Perhaps he did not have the vocabulary, or perhaps he simply did not see the need: it is as obvious that the swift wolves will prevail as that the biased wheel will tend to *red*. Either way, this silence makes it necessary to argue for my interpretation of Darwin's reasoning indirectly: I must hypothesize implicit, perhaps even unconscious, rationales to explain conclusions for which Darwin offers no argument. This method will of course be familiar to readers of part two.

Second, the inference is inherently probabilistic. Any sequence of outcomes is recognized as, in principle, possible: the wheel of fortune biased in favor of *red* might yield *black* every time, and the slower wolves might by pure chance do better than the faster ones. The sequences that are predicted—sequences in which *red* and swift wolves predominate—can be singled out only on the grounds that they are, in the long run, overwhelmingly more probable than the rest.

Third, the inference is inherently causal. It is by understanding aspects of the physical dynamics of the wheel of fortune and the braking pads, or of the die, or of the interactions between wolves and deer, that you reason your way to the conclusion that one outcome or variant will outdo the other. More exactly, it is the reasoning in step (1) of the inference that is causal.

Fourth, step (1) is also the inductively richest step. It is here that you convert nonstatistical, causal facts into facts about physical probabilities. It is here that the action is. And it is here that the equidynamics is. Step (1), then, is my focus in what follows.

Why was Darwin so confident that, in his hypothetical case, the swifter wolves would make superior deer hunters, stochastically speaking? Why did his readers share his confidence—even those readers who rejected other aspects of the *Origin?* Why does his claim seem equally reasonable to us today? So reasonable, in fact, that it may seem blockheaded to ask this very question: is it not obvious that a swifter wolf will on average do better? Deer are fast; wolves hunt by giving chase; faster wolves will chase more effectively. What could be simpler than that?

What could be simpler than seeing that the outcomes from a roll of a fair die are equiprobable and those from a loaded die not so much? But from chapter 6, you know that the dynamic reasoning required to deal with a shaken cube is not at all simple. It is, in fact, something of a marvel that we humans can predict anything about the outcomes of die rolls. We are so accomplished with probabilistic reasoning of this sort that we do not recognize the complexity of the problem that we solve when we see that a 5 is as likely as a 3. Likewise, in the biological case, we do not appreciate the complexity of the reasoning needed to see that swiftness and not, say, fur length, has an impact on a wolf's probability of capturing a deer.

To recover a suitable sense of wonder, compare the inferential prowess of Darwin and his readers to that of a race of eminent determinists, skilled in Laplace's demonic art of applying the fundamental laws of physics to particular sets of initial conditions to predict particular outcomes, but lacking heuristics for probabilistic reasoning. Ask these Laplacean predictors: which is more likely to catch a deer, a wolf made furrier or a wolf made swifter? They will reason as follows. Both swiftness and fur length can go either way. The swifter wolf might pass through a thicket ahead of its slower cousin, who arrives at just the same time as a juicy stag. The less furry wolf might, feeling the cold more, go out for a run on a winter's day to warm the blood, and run into the same stag—or a hungry cougar. To calculate the relative advantage of greater speed and greater fur length, then, it is necessary to calculate the relative frequencies with which these different scenarios occur.

The Laplaceans will therefore consider every possible situation in which a wolf and a deer are in close proximity. (Suppose for simplicity's sake that wolves are solitary hunters, and ignore the cougars.) They will calculate whether or not, in such a situation, the wolf will sight and catch the deer, with and without increments in speed and fur length. They will weight the different possible sets of initial conditions according to their historical frequency of instantiation. They will then compare the overall capture rate with and without increased swiftness and fur length, and give you your answer.

Or at least, that is their plan. They will, however, run into two considerable difficulties in executing the plan. The first and more obvious is that they will require immense computational powers, vast knowledge of the physical workings of the world, and even more patience. The second is that Darwin's specification of his imaginary scenario does not contain nearly enough detail to make the Laplacean calculations. Most notably, the predictors' knowledge of the exact details of the initial conditions of every actual earthly event will be useless to them in determining what I called in the previous paragraph the "historical frequency" with which the different possible initial conditions for wolf-deer encounters are instantiated, because the example is historically unreal. Yet we look at the same sketchy description and make an immediate judgment that increasing swiftness but not fur length will make the wolves better hunters.[9]

How do we do it? How do we cut through the bewilderingly dense causal complexity of a wolf-deer ecosystem, multiplied by decades or centuries of evolution? And how do we make what is intuitively the correct judgment when the situation is so egregiously underspecified? With only, evidently, the most meager mental effort?

Here is a clue. The Laplacean predictors would encounter the same obstacles in solving the following problem: twelve red balls and sixty white balls are sitting mixed up in a box. The box is shaken; a developmental psychologist then blindly draws five balls. Which is more likely to appear: a sample of four red balls and one white ball, or the reverse? First, the physics of even so simple a setup is formidable; to follow the paths of the balls around the box we earthlings would have to make fabulously precise measurements of initial conditions and then enlist banks of integrated circuits. Second, to a nonstatistical mind, the question is in any case underspecified: it supplies the Laplacean predictor with neither a frequency distribution over initial conditions nor enough information about the physical implementation of the setup—the size and weight and elasticity of the balls, the size of the box, the parameters governing the psychologist's selection of the balls—to calculate the outcome of any particular drawing.

Yet the problem is solved by six-month-old babies.

Darwin and his readers are clever in the same way. By introducing well-founded probability distributions, and then applying some general principles about the proportion of initial conditions resulting in one kind of outcome rather than another—that is, by applying proportional dynamic reasoning—they are able to think about wolves and deer in much the same way that they think about balls and urns, gliding over the tangled causal undercurrents and eddies responsible for individual lives and deaths. Or so I will suggest in the next two sections, where I develop a psychological model of the way in which, in the absence of statistics, normal humans reason probabilistically about fitness.

Two remarks before I continue. First, you might wonder whether there is not some nonprobabilistic way to infer a connection between the speed of wolves and their success in the chase, drawing conclusions about the frequencies with which different wolves capture their prey without using explicitly stochastic reasoning. You might wonder, that is, whether there is not some middle way between the onerous Laplacean task of calculating trajectory after possible trajectory, and the kind of reasoning that first assigns physical probabilities to various biologically significant events and then derives frequencies from those probabilities. Nothing I have said precludes the possibility of a form of nonstochastic biological intuition that solves the Laplacean problem in this way. I have no idea, however, in what way such reasoning might proceed, whereas I do see how the introduction of dynamic probabilities and the application of probabilistic reasoning allows an ordinary human reasoner to solve the problem—besides which, as already noted, Darwin's many references to chance and Wallace's invocation of the law of averages (note 7) indicate rather strongly that the pioneers of natural selection thought their way through the problem stochastically.

Second, it bears repeating that, in contrast to the case of the urn, our judgments about fitness are fallible and often highly qualified, at least when they are made about real as opposed to imaginary scenarios. The next chapter will explain how such judgments have played and why they continue to play, despite their limitations, an important role in evolutionary biology. For now, however, my focus is not on the scientific significance of the judgments but on their possibility; that is, on how we reach even prima facie conclusions about selective advantage in the face of the outrageous complexity of the biological world.

9.3 Still Life

Consider an example partway between balls and wolves, a streamlined version of a common predator-avoidance strategy. Some organisms, upon detecting the presence of a possible predator, stop moving and become entirely motionless. It seems obvious that, in the right sort of habitat, this "freeze" strategy will increase an organism's chances of remaining unnoticed by the predator. How do we reach this conclusion? What makes the advantage conferred by freezing so "obvious"?

Let me begin by specifying some further facts about a fictional habitat in which we are disposed to consider freezing to be advantageous, so as to paint more clearly the backdrop against which such judgments are made. Here and in what follows, my aim is simplicity rather than biological *verismo*.

Distinguish two kinds of predator-prey encounter, "distant" and "close" encounters. For a distant encounter, I will suppose, any set of initial conditions that leads to a stationary prey's being detected by the predator also leads to a moving prey's being detected, but not vice versa, so that there are some initial conditions where the predator will spot the prey just in case it keeps moving.[10] On the assumption that the conditions where freezing saves the prey do not have zero probability, then, it is easy to see that the freeze strategy increases the probability of the prey's survival.

Close encounters are where the insults and injustices of life in the wild are made manifest. In some close encounters, the freeze strategy will (let us say) backfire. Had the prey kept moving, the predator would not have seen it, perhaps because of favorable ground cover or perhaps simply because the predator is distracted. But because it freezes, the predator stumbles across it. The prey has stopped in the worst possible place, directly in the path of the approaching danger. You can add various other tragic scenarios of the same sort; all have the same property, that they are conditions in which the prey is detected just in case it stops moving. There are also, of course, many close encounters in which the opposite is true, that is, in which the prey is detected just in case it continues to move, as well as situations in which the prey is detected, or undetected, either way.

Two questions of evolutionary significance might be asked about the freeze strategy:

1. Is the strategy the sort of thing that would make a difference to an organism's chances of survival, and if so, why?
2. How much and what kind of difference does the strategy make? Does it increase or decrease the organism's chance of survival?

The first question is that of *whether* the freeze strategy is probabilistically relevant to survival; the second question, which presupposes an affirmative answer to the first, is of *how much and what kind* of relevance it has. There is a zeroeth question as well: is there a stable probabilistic fact of the matter concerning the strategy's relevance, or does its selective significance wax and wane as environmental background conditions fluctuate from day to day?

How do we answer such questions? In principle, you might proceed as follows: assume that there is some stable probability distribution over the initial conditions of predator-prey encounters (this implies an affirmative answer to the "zeroeth" question), and then calculate the ratio of sets of initial conditions where the strategy helps the prey relative to the status quo or default, in this case not freezing, to those where it hurts the prey, probabilistically weighted by the initial condition distribution and by the degree of hurt and help. If the ratio is even, then the strategy is selectively neutral. If it is greater than one, then the strategy promotes survival relative to the status quo; otherwise, it inhibits survival. What explains the strategy's relevance, if it is relevant, is whatever explains the ratio's differing from 1:1.

In practice, I suggest, our thinking about the freeze strategy loosely follows this ideal. We compare the ratio of sets of initial conditions where the strategy helps the prey to those where it hurts the prey, without probabilistic weighting.[11] If the ratio is far from even, we infer that the strategy is relevant to the prey's survival in the indicated way (and perhaps, if the ratio is even, we infer that the strategy is irrelevant). In the case of the freeze strategy, for example, we focus on close encounters in particular, and note that the situations in which freezing is disadvantageous require a rather special arrangement of initial conditions: the freeze has to place the prey on or very near the predator's precise trajectory. There are many more conditions in which freezing is advantageous compared to not freezing, so on the basis of close encounters alone, freezing is a selectively positive strategy, contributing to the prey's chances of survival. (Consideration of distant encounters cements the case.)

Under what circumstances will the comparison deliver reliable results? Under what circumstances, that is, will the ratio of probabilistically unweighted favorable to unfavorable initial conditions at least roughly track the ratio of probabilistically weighted initial conditions? Two conditions sufficient for rough tracking are

1. The initial conditions—such as the distance of predator and prey when the predator is first detected, the angle of one to the other, the ambient light, and so on—are smoothly (that is, microequiprobabilistically) distributed.[12]
2. The outcomes are sensitively dependent on initial conditions, so that the advantageous and disadvantageous sets of initial conditions are well mixed together in the initial condition space.

Then the disadvantageous conditions cannot be, proportionally speaking, much more heavily probabilistically weighted than the advantageous conditions, and so the probabilistically weighted ratio cannot depart too far from the unweighted ratio. It follows that when the unweighted ratio is high, the weighted ratio is at least greater than even.[13]

Call the rule implicit in this inferential practice—the rule that infers that a causal factor raises the probability of an outcome if its addition to the system in question diverts many more initial conditions toward that outcome than away from it—the *majority rule*. (Under that same heading might be included a closely related rule that infers from the preponderance of initial conditions producing an outcome that the outcome is more probable than not.)

Are the requirements of microequiprobability and initial condition sensitivity built into the majority rule? That is, does the rule impose the requirements as conditions of its application? I am inclined to answer in the affirmative, and I will proceed on this assumption, but the alternative is certainly possible: the rule requires only a comparison of ratios, and it earns its keep because the relevant initial conditions are in fact usually microequiprobable and the relevant dynamics are in fact usually sensitive to initial conditions.

Putting aside the question of initial condition microequiprobability, a reasoner employing the majority rule needs only a certain facility with proportional dynamics to make prima facie judgments about fitness—a certain facility, that is, with questions concerning the proportions of initial conditions leading to one kind of outcome as opposed to another. Our skill at proportional dynamic thinking has already been noted elsewhere: we find it fairly easy to see that allowing a tossed coin to bounce will not bias it toward or away from heads, or that an "army blanket roll" of a die will not yield all six outcomes with equal probability. This skill extends, apparently, to reasoning about animal behavior and related matters. We can easily see, in the example above, that the prey's freezing will be disadvantageous only in relatively few circumstances; equally, we can easily see that the prey's deviating 5 degrees to the left upon detecting a predator will be advantageous about as often as it is disadvantageous, from which we conclude that such behavior is unlikely to be selectively significant. Just as important, we can think intelligently about biological conditions under which a small deviation might be selectively significant, for example, if a predator on spotting the prey charges at it in an unwaveringly straight line.

Judgments about the selective advantage of strategies or other traits made with the majority rule are conditional, because the proportional dynamic thinking that underlies them is based on suppositions about the biological facts on the ground that are typically not known for certain to be true. Even in the case of the freeze strategy, we cannot be unconditionally sure that freezing is advantageous—the specification of the case above leaves open the possibility that, for example, the prey moves on so bare a landscape that, frozen, it is no less visible than when moving. Implicitly, then, we have in mind certain rough strictures limiting the scope of our judgment, and when applying the judgment to a real ecosystem, rather than to an imaginary scenario, we do so defeasibly, knowing that the implicit conditions may not be satisfied.

What about other fitness comparisons? Suppose a deer is in the vicinity of a swift wolf. Does the wolf's swiftness increase the probability of its seeing and catching the deer? The majority rule instructs us to compare the initial conditions of such scenarios with and without swiftness. If adding swiftness to the mix means many more unfavorable initial conditions are made favorable (the swift wolf sights and captures the deer where a slow wolf would not) than favorable conditions are made unfavorable (a slow wolf would have succeeded where the swift wolf does not), then swiftness confers a net selective benefit.

You can easily see that this condition holds. There are many cases where only the swift wolf, in virtue of its swiftness, catches its prey. There are a few where swiftness prevents its seeing the deer (it is in the wrong place at the wrong time), but these are balanced by those where slowness prevents a slow wolf seeing the deer for parallel reasons. Finally, there may be a few conditions where swiftness is disadvantageous in the chase, because a swift wolf experiences some mishap that it would have avoided were it traveling slightly more slowly, but these are presumably relatively few in number. The net effect of extra speed, then, is to increase the probability that the wolf takes the deer.

The Darwinian judgment about the advantage of swiftness in the pursuit of deer, like all such judgments, comes with implicit conditions attached—for example, that wolves and deer are similarly fast, so that a little extra speed makes a difference. (Additional speed would likely be of no help to a wolf that preys on sloths, whose top speed along the ground is at most 2 kilometers per hour.) Further, comparative fitness judgments of the sort discussed here take no account of the energetic or other structural costs of a trait. As techniques for predicting what will and will not be selected, they are therefore highly provisional. But simply by showing how we can perceive selective tendencies in an ecosystem's dynamic tangle, the majority rule provides the beginnings of an explanation of our ability to overcome biological complexity and underspecification, and so to outthink Laplace's demon determinist, when making fitness comparisons of the sort that provide the backbone of Darwin's *Origin*.

9.4 Biological Microequiprobability

The majority rule requires its users, I suggested cautiously in the previous section, to have some reason for believing that the relevant initial conditions—predator position, prey position, and so on—are smoothly, or microequiprobabilistically, distributed. A microequiprobability condition of this sort is also required for the biological application of the equilibrium rule, to be discussed in the next section. Let me therefore pause to ask where, in equidynamics, information about the distribution of biological initial conditions might originate.

I see two possible sources. On the one hand, we might treat the initial conditions as exogenous, meaning that we do not actively examine the process that generates them, rather defeasibly supposing that they are, like other environmental "noise," microequiprobabilistically distributed. An equidynamic rule warranting this supposition, which I call the microequiprobability rule, will be developed in section 12.1. The microequiprobability rule applies to almost any kind of initial condition. A more narrowly targeted version of the rule—in section 12.2 it is called the physiological microequiprobability rule—might warrant our treating the position, orientation, and perhaps some other properties of mobile organisms in particular as microequiprobabilistically distributed, without any further knowledge of such organisms' internal dynamics.

On the other hand, we might derive the microequiprobability of biological initial conditions from the microdynamic rule (section 7.1). The biological application of that rule is the main topic of this section.

For an encounter between predator and prey, the microdynamic rule would be used as follows:

1. Assemble the various kinds of biological processes that determine the initial conditions of a predator-prey interaction: earlier predator-prey interactions, foraging decisions (e.g., determining in which area to search for food), other foraging behavior (e.g., patterns of searching within a given area), and so on. For the sake of continuity, I call all such processes "interactions," although they may involve only a single organism interacting with its environment.
2. For each of these interactions, verify sensitivity to initial conditions.
3. For each of these interactions, verify the smoothness, that is the microlinearity, of the dynamics.

Given sensitivity and smoothness, the microdynamic rule allows you to infer the existence of a microequiprobabilistic physical probability distribution over the post-interaction state variables, such as predator position and orientation, along with the independence properties that microequiprobability implies.

Can the microdynamic rule realistically be applied to biological dynamics? It might seem not: there are invariably some aspects of an organism's post-interaction state that do not depend sensitively on its pre-interaction state. An organism's weight, for example, might change in a small way as a result of a light snack, but the degree of dependence is minimal.

To apply the microdynamic rule to such cases, divide the organism's state variables into those that depend sensitively on the pre-interaction state and those that do not—into what I will call the sensitive and the insensitive state variables. Then individuate interactions so that a difference in the value of an insensitive state variable is sufficient for a difference in interaction type. Interactions that are identical except for the organism's weight, for example, will according to this criterion count as different kinds of encounter. In effect, the insensitive variables are no longer treated as variables at all.

Since the microdynamic rule requires that the sensitivity and smoothness tests be passed for every relevant kind of interaction, the effect of this move is as follows: the microdynamic rule can be applied, and a microequiprobabilistic distribution over the sensitive state variables inferred, provided that for every possible assignment of values to the insensitive variables, the post-interaction values of the sensitive variables depend sensitively and smoothly on their pre-interaction values.

Two remarks about this approach. First, it licenses no conclusions about the probability distribution over the insensitive variables. Second, although in principle it dictates infinitely many tests in order for the microdynamic rule to apply, in practice these tests can be amalgamated into a few calculations. I can see, for example, that for any value of weight, an organism's post-encounter orientation depends sensitively on its pre-encounter orientation without having to consider individually every possible weight. The approach, then, is not impractical.

A new problem: what if *no* post-interaction state variable depends sensitively on the pre-interaction state? Suppose, for example, that you wish to apply the microdynamic rule to an organism that forages using a fixed search pattern (for example: divide the area into rows, and then start at the top row and work your way down, traversing each row from left to right). Arguably, neither position nor orientation depend sensitively on their earlier values for this particular kind of "interaction"; since the microdynamic rule requires that every interaction pass the sensitivity and smoothness tests, it might seem that the rule cannot be applied to such a creature at all.

The solution is to redescribe the organism's state using a different (though equally standard) set of variables, so that those aspects of the state that do depend sensitively on their earlier values—and I speculate that for most mobile organisms, there are always some such aspects— have their own dedicated variables.

Two examples. First, a parameter such as position may be divided into coarse-grained position—that is, approximate position—and fine-grained position. You represent an organism's position, then, using two variables, one of which tells you approximately where the organism is, that is, what region it occupies, and the other of which tells you exactly where in that region it sits.[14] (Such a division is already implicit, I note, in the notion of microequiprobability, which is uniformity over a variable's fine-grained component.)

Second, and more complexly, a parameter may be divided into a relative and absolute component, as when the velocity of two particles is divided into, first, their velocity relative to one another, and second, the velocity of their center of mass. In the biological case, this may require reconceiving an entire ecosystem in terms of the relative positions of predators and prey.

When things go well, the aspects of a system's variation will be separated using this method into sensitive and insensitive variables; the microdynamic rule can then be used to derive the microequiprobability of the sensitive variables. This is only useful if the sensitive variables can then be used in a straightforward way in proportional dynamic reasoning; fortunately, in both examples this seems likely, as the parameters of a predator-prey encounter, say, are often conceived of in terms of relative or fine-grained position—indeed, such a parameterization may render the proportional dynamics easier than the more obvious parameterization in terms of absolute position.

Do we actually use these tricks in applying the microdynamic rule? I am not sure. Perhaps we use other shortcuts, or perhaps the rule comes in several specific versions, each suitable for particular circumstances. Given the paucity of empirical evidence about our equidynamic reasoning, it is surely best to keep an open mind; nevertheless, progress is most likely to be made, I believe, by formulating and testing specific versions of putative equidynamic rules.

9.5 Sweet Flowers

Darwin's second example of natural selection concerns the coevolution of nectar-bearing flowering plants and the pollinating insects that feed on them. He begins with the plants:

> Certain plants excrete a sweet juice, apparently for the sake of eliminating something injurious from their sap: this is effected by glands at the base of the stipules in some Leguminosae, and at the back of the leaf of the common laurel. This juice, though small in quantity, is greedily sought by insects. Let us now suppose a little sweet juice or nectar to be excreted by the inner bases of the petals of a flower. In this case insects in seeking the nectar would get dusted with pollen, and would certainly often transport the pollen from one flower to the stigma of another flower. The flowers of two distinct individuals of the same species would thus get crossed; and the act of crossing, we have good reason to believe . . . would produce very vigorous seedlings, which consequently would have the best chance of flourishing and surviving. Some of these seedlings would probably inherit the nectar-excreting power. Those in individual flowers which had the largest glands or nectaries, and which excreted most nectar, would be oftenest visited by insects, and would be oftenest crossed; and so in the long-run would gain the upper hand. (Darwin 1859, 91–92)

Darwin then goes on to discuss the evolution of advantageous stamen and pistil placement, of dioecy (separate male and female plants), and of the physiology of the pollinating insects, but I will focus on the conjectured explanation of the evolution of nectar-secreting glands (nectaries) within the flower.

Consider the evolutionary stage described in the passage's final sentence, in which all flowers contain nectaries, but some secrete more nectar than others. Darwin claims—and as with the wolves, the claim seems entirely reasonable—that the more abundantly nectariferous flowers will be visited more often (or, he might perhaps have added, for longer) by nectar-seeking insects. Consequently, the pollen of such plants will be spread more widely, and the plants themselves will receive pollen from further afield. On the assumption that outcrossing on the whole produces more "vigorous" offspring, the most generous providers of nectar will have the most successful descendants. (As in Darwin's other examples, the energetic and structural costs of nectar manufacture are for simplicity's sake ignored.)

Each of the steps in this explanation of the advantages of abundant nectar provision is stochastic; the advantage accrues to the most generous providers only "on average" or "in the long term." Darwin is explicit about the last stage: the outcrossed and therefore most vigorous seedlings have the "best chance" of flourishing and surviving. But as his talk of the "long run" indicates, and as his frequent references to the uncertainties of life elsewhere show, he would have been keenly aware that things could go wrong for the most fecund nectar producer at any point.

Let me count the ways. The insects that visit a flower might bring no pollen from far away, if they arrive first thing in the morning, or if they have been browsing other species. Or the pollen they take from that flower might never reach a suitable destination, if it is the end of the day, or if they go on to browse other species, or if they are taken by predators or otherwise struck down on their way to the next nectar source. Even if they do reach another suitable blossom, they might happen not to brush against the carpels in quite the right way to transfer the pollen, or they might transfer another flower's pollen instead.

Or perhaps no insect visits a prodigious nectar producer at all. This might be through short-term or long-term bad luck. Bad luck in the long term is a plant's happening to grow in a place seldom frequented by insects; bad luck in the short term is a perfectly well-situated plant's simply happening not to lie on the flight path taken by the local insects that day.

Despite these many possible mishaps, we follow Darwin in judging instinctively that a flower's ability both to attract pollen from far away, or to send its own pollen in return, or both if it is monoclinous (hermaphroditic), will on average increase with the quantity of nectar that it provides. By this "easy" insight we discern order in a thick web of biological complexity. How? Equidynamically—as I will show by examining the kinds of inferences we produce to argue for the unimportance, in the long term, of the misfortunes described in the preceding paragraphs.

First, consider the possibility that prolific nectar producers, by contrast with average producers, mostly grow in locations that are rarely visited by insects. Were this a long-term trend, there would be no advantage to improved nectar production. We have reason to think that there is no such trend, however: the seeds of improved producers will disperse in just the same way as the seeds of regular producers; thus, although there is some probability that the improved producers will consistently land in worse locations than the regular producers, it is very small. In thinking about seed dispersal, we think physically: the seeds are much like molecules of a gas (if imagined drifting on the wind), or balls plucked from an urn (if imagined snagged on the coat of some shaggy animal). There is no need, here, for a distinctively biological approach to reasoning about the relevant probabilities.

Second, consider the possibility that insects that have recently visited a prolific nectar producer are captured by predators in proportionally greater numbers than visitors to regular producers. Again, we have reason to think that the probability of such a trend continuing for long is minute, and again the central premise is that the probability of a pollinating insect's being eaten is the same whether it has visited an improved nectar producer or not. A direct analogy to physical systems cannot supply this premise, however. We must find some biological basis for drawing probabilistic conclusions about the movements of predators and prey.

What we do with the prey, the pollinating insects, is perhaps something roughly as follows. We infer a microequiprobabilistic distribution (or distributions) over the organisms' position and orientation using the microdynamic rule, as explained in the previous section. We treat "microposition" and "micro-orientation" as stochastic variables in order to apply the equilibrium rule, assuming that changes in a browsing insect's state depend sensitively on small differences in its earlier facing and place (section 7.2). From the equilibrium rule, we then infer the existence of a unique equilibrium distribution over the insects' position within whatever patch they are browsing (with a short relaxation period).

In the case at hand, there is no need to deduce a particular form for the distribution; what matters is that the distribution has the same form for visitors to improved flowers as for visitors to regular flowers. Why the same? Because the equilibrium rule's equilibrium distributions are entirely determined—so probabilistic dynamics tells us—by the parameters of the random walk induced by the stochastic variables, and these parameters are unrelated to the only property that distinguishes the two classes of visitors, namely, the size of the just-visited flower's nectar bounty.

Since the visitors to improved flowers are distributed—with respect to position, orientation, and whatever else makes a difference to one insect's being eaten rather than another—in the same way as visitors to regular flowers in the same patch, they will have the same chance of being taken by predators. Identical considerations justify our belief that a visitor to an improved flower is just as likely to be carrying faraway pollen as a visitor to a regular flower.

We reason, I should add, with our eyes open: we can appreciate the conditions under which our inferences will fall through. While we assume, for example, that visitors to improved producers are no more likely to be captured by predators than visitors to regular producers, we are also aware of the conditions under which this assumption will fail, as would happen, for example, if the visitors to improved flowers indulge themselves to the point that they are overladen, and therefore slow, conspicuous, and easy to catch. Were this tendency to reckless gluttony to exist, we would no longer consider improved nectar production to be an obviously advantageous strategy.

Third, why do we conclude that an improved flower will be more frequently visited than an equally well-situated regular flower? Suppose a group of insects is browsing a mixed field of flowers, some with improved and some with regular nectar production. The insects are moving among the flowers in that haphazard way that insects do. Using the equilibrium rule as above, followed by the uniformity rule, we can infer that on average, the insects are evenly distributed over the field. (The application of the uniformity rule requires some further assumptions about the short-term probabilistic microdynamics of foraging insects' movements: roughly, that their directions of travel as they move around the field are themselves uniformly distributed.)

Consider a particular insect in a particular patch of the field. Within that patch will be a handful of flowers. Suppose that the insect has the ability to detect, at a short distance, the difference between an improved and a regular flower, and that it prefers improved flowers. Then it will tend to forage in the improved rather than regular flowers in its patch.[15] The same goes for its conspecifics uniformly distributed across the whole field. The improved flowers will therefore receive, on average and in the long run, a greater number of visits than the regular flowers.

You might reach a similar conclusion by way of slightly different assumptions about the foraging life of the insects; for example, you might suppose that the insects cannot differentiate improved and regular blossoms from a distance, and so visit flowers totally at random, but that they feed for longer at improved flowers, thereby receiving a more thorough dusting of the improved flowers' pollen. Since insects are distributed uniformly in a patch, all flowers will on average receive the same number of visits; because each such visit takes more pollen from an improved than from a regular flower, the improved flowers on average transfer more pollen to other parts of the system (and the same goes for the receipt of pollen from other parts). What matters for the study of biological equidynamics are not the details of the reasoning, which might vary from person to person, but the kinds of cognitive resources that supply its probabilistic premises, which are much the same on either version.

How do these various considerations come together in a typical reader's grasp of Darwin's hypothetical explanation of the evolution of nectariferous flowers? The most plausible story, I think, is roughly as follows. When we read Darwin's description, we construct for ourselves a simple model of the ecological system much like the model laid out above, with bees randomly buzzing around a field of flowers from patch to patch. (It is possible that some empirical information is used to configure the scenario; for example, our observations of bees' haphazard browsing might prompt us to represent their foraging as a stochastic process rather than as a regimented deterministic search.) We then draw a provisional conclusion from the model, in this case that the bees preferentially visit the improved flowers.

A question is then posed: is the simple model relevantly similar to more realistic fleshings-out or variations of the scenario? Can the same conclusions about selective advantage be drawn for the same reasons if, say, the fields of flowers are irregularly shaped, or if the improved and regular flowers are not mixed together but tend to cluster each in their own patches, or if the pollinating insects retain some memory of the locations of the most bountiful flowers, or if the insects mark flowers to indicate that they have been visited recently? To the extent that we obtain affirmative answers to these questions, our confidence grows that the reasoning applied to the simple model, and so the conclusion about selective advantage based on that reasoning, generalizes.

We might go even further in investigating the scope and limits of the simple model. For example, we might, at least if certain questions are made salient, go through the explicit reasoning needed to conclude that improved flowers are on average no better or worse situated than regular flowers, or that insects visiting improved flowers are no more likely to be eaten immediately afterward than visitors to regular flowers. (In addition, we thereby come to see the circumstances under which these assumptions would fail, as shown in the discussion of the overladen bees earlier in this section). In this way, our reasoning about the wolves, the flowers, and other such scenarios may become quite involved and sophisticated; however, we can reach a prima facie judgment early and quickly by taking just the first step.

A part of the rhetorical power of Darwin's scenario-building subsists in this flexibility, I think: readers who are sympathetic or rushed can quickly grasp the essence of Darwin's point, though at the cost of some epistemic security, while more cautious or skeptical readers can pause to reflect on the biological complexities of likely real-life implementations of the scenario. These latter readers, rather than throwing up their hands at the complications, find that equidynamics gives them the guidance they need to appreciate the considerable reach of Darwin's claims, and also the tools to grasp the many implicit riders that those claims ought to carry.

The story so far: Deciding whether a strategy or trait is selectively significant, and if so, why and how, is a dastardly problem, given the complexities of biological dynamics. Yet it is a problem that we as a species appear to have solved, so comfortably that questions about the advantages of swiftness in wolves, the freeze strategy, and so on, seem to us to be entirely straightforward.

We do not answer these questions by consulting the relevant statistics. Most of us have little experience of the biological world. In any case, statistics for many such questions are not available to any of us: the habitats in question long ago ceased to exist, or in the case of hypothetical habitats, have never existed. So we answer the questions using the same equidynamic techniques that we apply to gambling setups, and that Maxwell applied to gases.

10

APPLIED BIOEQUIDYNAMICS

10.1 The Equidynamic Origin of Evolutionary Theory

One long argument—so Darwin aptly described the structure of *On the Origin of Species*. The premises of the argument are many, but one is perhaps more important than any of the rest. In order to convince his readers of the viability of the gradualist theory of evolution by natural selection, Darwin needed to persuade them that small differences in phenotype could and frequently did have long-term evolutionary consequences, with the more "adapted" phenotype most likely going to fixation at the expense of its slightly less adapted competitors.

He presses this point again and again in his discussion of natural selection: "any advantage, however slight" suffices for selection (1859, p. 81); "the slightest difference of structure or constitution may well turn the nicely-balanced scale in the struggle of life, and so be preserved"; "characters and structures, which we are apt to consider as of very trifling importance, may thus be acted on"; and, of course, "natural selection is daily and hourly scrutinizing, throughout the world, every variation, even the slightest; rejecting that which is bad, preserving and adding up all that is good" (all from p. 84).

The argument for this biological lemma has two premises. The first is that small differences in probability, and in particular small differences in expected numbers of offspring between two variants, can make big differences to the variants' numbers in the long run, ultimately likely bringing about the extinction of all but the most productive type. Darwin provides no mathematical argument for this proposition; it comes presumably from his intuitive gamblers' grasp of the law of large numbers.[1]

The second premise—or perhaps it is more of a presupposition than a premise—is that small differences in phenotype will at least occasionally manifest themselves as persistent, consistent differences in expected offspring number. To Darwin it is clear that "variations useful in some way to each being in the great and complex battle of life should sometimes occur in the course of thousands of generations" (80). He would appear to be thinking as follows: either a variation is useful—it increases an organism's expected reproductive output—or it is not useful. Given enough time, surely some variations at least will fall into the former category?

The argument turns on a false dichotomy, however—or rather, it succeeds only in the context of a wildly strong background assumption, that there is typically a fact of the matter as to whether a variation increases, or decreases, or leaves unchanged an organism's expected reproductive output. How could there be no such fact? Quite easily: there may be no such thing as *the* expected offspring number for an organism, because there is no such thing as *the* probability distribution over offspring number, nor anything like it.

And how could *that* be? A definite probability distribution could fail to exist for two reasons. First, it might be that there is no physical probability distribution at all over offspring number. Second, it might be that there is such a distribution, but that it is sensitive to transitory environmental conditions, so that the shape of the distribution changes from week to week, and perhaps day to day or hour to hour, as the environment changes. (Suppose that the environmental changes do not themselves exhibit the sort of patterns that can be captured by a context-independent probability distribution.)

Let me focus on the second possibility. If an organism's reproductive distribution varies with the hour, then a trait's contribution to the probabilities may vary in the same way: at one time it might put the organism at an advantage relative to its rivals, while at another time it might drag the organism down. Because the environmental conditions change nonprobabilistically, there would be no statistical pattern to the shifting of the ecological balance for and against such a creature; thus, the long-term advantage or otherwise of the trait would depend on nothing less fine-grained than the actual facts about low-level environmental variation over the creature's lifetime.

Under such conditions there would be no general tendency for one trait to be selected at the expense of another. At one time, this trait would have the upper hand; at another time, that trait. The selective pendulum would swing back and forth like the damped, driven pendulum that canonically exemplifies classical chaos, moving with no discernible rhythm. Such a process might maintain the balance between different variants in a population indefinitely, but it is perhaps more likely (in a nonphysically probabilistic sense of "likely") that, one by one, variants would happen to experience a run of bad luck leading to extinction, just as with genetic drift in small populations, but without the concomitant statistical regularity. One trait would be left, but there would be no sense in which it survived because of its inherent ecological superiority. (Extremely maladaptive traits—for example, developmental variations that ensure the death of the organism before sexual maturity—would of course be selected against; my focus is on competing traits that exhibit Darwin's "slight differences.") Thus, the probabilistic properties of evolutionary change would not explain evolved organisms' adaptedness to their habitats, and plausibly, if this is how evolution worked, organisms would not be particularly well adapted, if there were any evolved organisms at all.

Why think that the reproductive probability distribution is sensitive to small-scale features of the environment? Darwin himself gives the reason, when he observes "how infinitely complex and close-fitting are the mutual relations of all organic beings to each other and to their physical conditions of life" (80). He is contending, in this passage, that even small phenotypic variations are likely to be selectively relevant. But the argument is dangerously strong: small things are relevant to selection, so the probability distribution will depend on many, many small things, a good proportion of which will come and go, or fluctuate in their intensity, over short intervals of time. To put it another way, "complex, close-fitting relations" are precisely what creates the sensitivity to initial conditions that is the core of chaos.

A real example of the sort of selective irregularity that Darwin's argument might lead you to expect is provided by the seed-eating Galápagos finches. In drought years in the Galápagos Islands, larger beak sizes are selected for; in flood years, smaller beak sizes. The average beak size therefore varies upward and downward from year to year with meteorological fitfulness; there is no such thing as the "most adaptive size" for a beak (Grant 1986). We can use Darwinian models to understand phenomena such as these, but if all evolutionary change were of this sort, Darwin's grand gradualist explanation of speciation and more generally of the adaptedness of life would utterly fail.

Happily for the *Origin,* the instability of finch beak size appears to be more exception than rule: most traits' contribution to the reproductive probability distribution is not entrained to short-term environmental variation, or at least—as with frequency-dependent selection—if it does depend on fast-changing aspects of the environment, the dependence is on at most one or two high-level variables. More extensionally, when we are able to measure organisms' reproductive success we usually find regular statistical trends; differences in rates of reproduction with and without the trait of interest stay fairly constant, and the fluctuations in such differences are of just the sort you expect in a simple stochastic process, much like the fluctuations in the frequency of heads in a long series of coin tosses. Darwin's implicit second premise is therefore often enough true.

Two questions, then. First, why is the premise correct? Why, despite the potential for reproductive chaos, are there sufficiently many stable, longish-term reproductive trends to bring about what Darwin wanted to explain, the fact that almost every ecological niche is inhabited by a species marvelously adapted to precisely those environmental conditions? The question of macrolevel ecological stability is a deep and fascinating problem; I try to give the outline of an answer in Strevens (2003).

It is the second question that finds its proper place in the present book: why, if Darwin recognized the sensitivity of ecological outcomes to fine-grained initial conditions, was he so sanguine about the existence of reproductive probability distributions sufficiently independent of low-level environmental change to drive long-term selective trends? And why were his readers on the whole willing to go along with him, on this matter at least?

Equidynamics is the answer. The techniques for comparing fitness discussed above are also techniques for determining what does and does not make a difference to the probabilities of various selectively significant events, such as predation, and indeed, for determining that there exists a physical probability distribution over such outcomes at all. More specifically, the equilibrium rule package can be put to use quite generally to infer the existence of probabilities for selectively significant outcomes which probabilities are mostly or entirely independent of low-level environmental conditions, giving Darwin the premise he needs to show that natural selection has the power to climb hills toward maximum adaptedness—the premise that establishes, in effect, that there exist such hills, or in other words that there exist gradients of fitness, for the climber to ascend.

Not just the fitness comparisons in Darwin's wolf and pollination scenarios, then, but the viability of the entire theory of evolution by natural selection hinges on an insight that, in the context of mid-nineteenth-century natural history, only equidynamics was in a position to provide.[2] This insight of sweeping stochastic independence from transient environmental change concerns, not the form of some master probability distribution such as Maxwell's velocity curve, but the master form for a vast range of probability distributions, those describing the probabilities of selectively significant outcomes for every mutant, every variant, every trait, that had ever existed, or would ever exist, on earth.

Just about all of this has been invisible to historians and philosophers of evolutionary theory. It has been invisible that Darwin's theory requires such a premise and that Darwin and his readers were willing to tacitly grant the premise in spite of the manifold tightly interwoven, apparently chaos-creating individual-environment relationships that determine organisms' ecological fortunes and misfortunes. Thus the role of equidynamics itself has been ignored. What comes easily comes quietly: we wield equidynamics to cut through biological complexity so adeptly that we barely notice what we have achieved, and we fail to notice at all the strength of the method used to achieve it.

10.2 Modern Times

Darwin's principal aim in the *Origin* is to show that natural selection is a plausible mechanism for explaining the sort of radical phenotypic change in virtue of which organisms become in an intuitive sense better adapted to their environments. His focus is therefore less on specific cases of selection, and more on the ubiquity and power of the mechanism in general.

Modern evolutionary biologists appealing to natural selection need no longer stake a place for selection as an evolutionary agent. Their work typically concerns either the explanation of particular evolutionary episodes—the development of melanism in moths, antibiotic resistance in *Staphylococcus aureus,* changes in beak morphology in Galápagos finches—or the evolution of types of physiological or psychological mechanisms, such as the mechanisms underlying signaling or altruistic behavior. In such investigations, what use is equidynamics?

Equidynamics alone cannot establish the correctness of a particular evolutionary explanation, as Darwin was quite aware. (In what follows, by "evolutionary explanation" I mean "explanation that invokes natural selection to account for evolution," though see note 4.) First, equidynamic judgments come with riders (e.g., section 9.5's "the bee does not feed so much that it weighs itself down"), and it takes empirical work to determine that the riders hold. Second, equidynamic judgments typically do not take into account metabolic, structural, and other indirect costs of traits. Third, equidynamics cannot tell you in what way, if any, a trait is passed on from parent to offspring. When using equidynamics to predict evolutionary trends in the long term, there are further obstacles, such as developmental constraints on phenotypic change.

What is equidynamics good for, then? Endler (1986, chap. 3) includes in his survey of "methods for the detection of natural selection in the wild" techniques that attempt to predict evolutionary change, that is, change in trait frequencies, from the "known properties of the traits, the environment, or both" (pp. 86–93). For the most part, the methods that Endler examines under this heading make no use of statistics—at least, no use of statistics concerning relative fitness—but rather proceed by constructing models of trait functioning which are used either

1. to determine which of several traits assumed to be present in the population at a given time best promotes survival or reproduction, and then to predict that or explain why that trait increases its representation in the population over time, or
2. to determine which of a range of traits allowed by the model best promotes survival or reproduction, and then to predict that or explain why that trait goes to fixation.

The first method predicts or explains evolutionary change in the population at a given time; the second method—which uses what are called optimality models to explain such things as foraging strategies—predicts or explains the end point of evolution, that is, the state of the system when it reaches equilibrium. It should be clear that, despite their differences, both methods call upon the same kind of nonstatistical reasoning to determine selective advantage, reasoning of a sort exemplified by the analysis above of ordinary humans' thinking about wolves, the freeze strategy, and nectar production.

Such inferences turn on judgments about physical probability that are guided, I have argued, by equidynamics. Even when the models in question do not represent physical probabilities explicitly, their validity turns on a kind of probabilistic thinking: because almost any trait can "back-fire" (harming rather than helping its possessor), and because almost every trait requires for its proper functioning that its possessor's ecological encounters are at least roughly statistically representative (that bountiful nectar-producers are visited by insects that carry roughly the average amount of pollen, for example), the advantage conferred by a trait is almost always an advantage "on average," "in the long run." Deterministic models simply take for granted, and render tacit, the assumption that organisms with the trait will experience statistically representative conditions on average and in the long run.[3]

After sketching several examples demonstrating the successes and limitations of statistics-free, ecological reasoning about optimality, Endler concludes by quoting Oster and Wilson (1978):

> Optimization models are a method for organizing empirical evidence, making educated guesses as to how evolution might have proceeded, and suggesting avenues for further research.

That seems to me to be a better place to start than to end. What is it about equidynamic thinking that renders it useful in these ways? Given that equidynamic reasoning typically cannot bear the burden of predicting or explaining evolutionary change unaided, how does it help to "organize evidence," "make educated guesses," "suggest avenues for further research"?

The answer, I propose, lies in equidynamics' behind-the-scenes contributions to the process of model building: when a model of evolution by natural selection is constructed, equidynamic thinking gives the model-builders reliable clues as to what should be included in the model from the beginning, what might be later added to the model if circumstances demand it, and what can safely be ignored entirely.[4] (The same can of course be said concerning the importance of physical, biological, and social intuition for model building across the sciences, whether that intuition is stochastic, deterministic, or something else—as already noted in my preface. Although in what follows I focus for obvious reasons on that branch of physical intuition that I call equidynamics, I do not mean to diminish the importance of physical intuition of all kinds in the model-building process, whether in evolutionary theory or elsewhere.)

Suppose, then, that you are building an evolutionary model, or better, a series of such models. You plan to construct a prototypical version, test it against the empirical facts, and then augment the model as necessary. What goes into the prototype? If you include everything that might make a selectively significant difference—whatever might cause a death here, prevent a death there—you will include everything. Your model will be a Laplacean monstrosity, capable perhaps in principle of predicting the course of evolution but in practice far too unwieldy to do real scientific work. Simplify, simplify, simplify—but not too much. That is the modeler's code. As a piece of practical advice it is useless, however, unless you know what to leave out and what to keep in.

Enter probabilistic thinking. The course of evolution by natural selection is determined by aggregate numbers of deaths and births: what matters is how many of one variant reproduce as opposed to how many of the others, not which particular organisms live, die, or are born. Thus, if we are justified in thinking of ecological processes as stochastic—if we are justified in supposing that there are physical probability distributions over ecologically important outcomes—then the true difference-makers are the factors that influence the probability distributions over the important outcomes, that is, that influence the probabilities of birth, death, and the rest.[5]

It is these factors that ought to be included in an evolutionary model. But not all at once. You should start by building a model including only those factors that are known to make a significant difference to the probabilities. In reserve you hold factors that might make a difference, depending on how things stand with certain as-yet unknown ecological matters of fact; also held in reserve for fine-tuning are those factors that make only a relatively small difference to the probabilities. (I say *relatively* because in a gradualist evolutionary scenario, the aggregate differences between the probability distributions for competing variants will themselves be small.)

To understand the evolution of nectar production in flowers, for example, you begin with as simple a model as possible, including only the factors that you are reasonably sure (if your evolutionary hypothesis is on the right track at all) make a difference to the course of natural selection, such as the amount of nectar produced by a flower, along with whatever assumptions are needed to derive from these factors the selective impact in question, for example, the assumption that insects on average distribute themselves uniformly over a patch of flowers, so that they find improved nectar producers as easily as they find regular nectar producers.

You then subject your model, as best you can, to empirical test. You might test it against actual evolutionary trends, empirically measured, to see whether the traits that the model declares to be advantageous are in fact advantageous. Or you might test the model by comparing the physiological and ecological assumptions made in the model to the actual system that the model is supposed to represent. As an example of this latter sort of test, consider the case of the greedy bees (section 9.5). The equidynamic conclusion that bees visiting improved flowers (flowers that provide more nectar than normal) are no more likely to be eaten than bees visiting regular flowers depends in part on the assumption that recent visitors to improved flowers do not so gorge themselves on nectar that they make themselves easy targets for predators. This supposition might be verified by observing the flight patterns of bees visiting nectar-rich flowers, or by constructing an artificial feeder that provides varying amounts of nectar and watching for any correlation between nectar provision and in-flight vulnerability.

If a model fails such a test, it must be amended to represent some previously ignored aspect of the habitat, or to represent some aspect more accurately. Here you call on one or more of your reserve variables, whose relevance was originally unknown but is now strongly suspected. At the same time, you disregard variables such as the transient positions of individual bees whose probabilistic relevance can be ruled out in advance, perhaps on the grounds that it is washed out by random walks, even if the facts about microposition can and often do make a difference between the individual's life and death.[6]

To build your model in this cumulative way, then, you must know, or at least have well-founded beliefs as to, which elements of your target ecosystem impact the probabilities of which selectively significant events. Such questions can in some cases be answered by gathering statistics. But to obtain the needed numbers is often too difficult or too expensive, or in the majority of cases where the ecosystems and variants in question have long ago vanished, impossible. The solution lies in equidynamics, with its manifold judgments of relevance and irrelevance based not on statistics but on the physical or biological constitution of organisms and their habitats.

Three qualifications. First, equidynamic judgments about probabilistic relevance are not a priori. Like judgments about die rolls and urn-drawings, they are based on information about the workings of an organism, its habitat, and the interactions between the two. This information may be conjectural, but insofar as it exists at all, it is based on empirical observation.

Second, equidynamics provides the judgments of probabilistic relevance that give evolutionary biologists a solid foundation on which to begin the building of models—to begin, but not, as I have emphasized above, to end. If at all possible, every one of a model's assumptions should be tested against, and refined in the light of, statistics about differential reproduction and of course information about whatever trait or equilibrium mix of traits has been fixed by selection.

Third, equidynamic reasoning is not the only way to kick off the model-building process; it is merely the best way. You might get started by guessing wildly or picking at random when deciding which variables to include in a model. Or you could adopt the "Laplacean" strategy suggested above: include everything relevant to individual births and deaths. But the only sensible rival to the equidynamic approach is one mentioned briefly by Endler (p. 88): appeal to broad rules of thumb about probabilistic relevance—generalizations of the form, "This factor tends to be probabilistically relevant to (say) death in such and such circumstances"—or, what is more or less equivalent, use the same factors in your present model that have worked for you and your friends in other models. Brute induction of this sort might indeed be useful on occasion, but I doubt that it yields information about relevance of either the scope or the subtlety of that supplied by equidynamics.

To summarize, equidynamics makes modeling possible in practice when statistics are expensive or unavailable by dividing potentially relevant properties, on principled grounds, into several classes:

1. Those properties that are almost surely probabilistically relevant (e.g., the number of bees that visit a given kind of flower). This class can be further divided into properties that have a major impact and those that have only a relatively minor impact on the probabilities in question.
2. Those properties the fact of whose relevance depends on as yet unknown facts about a system (e.g., the amount of nectar ingested by a bee when feeding on a given kind of flower).
3. Those properties that are almost surely probabilistically irrelevant, such as the moment-to-moment positions and orientations of individual organisms.

4. Those properties whose relevance is not treated by equidynamic reasoning about organism-environment interactions. Most notable are such internal matters as metabolic cost and inheritance.

Advice to the model-builder: properties of the first sort should appear in even the simplest models of the phenomena; properties of the second sort are on standby, meaning that they may need to be added to the model if empirical investigation demands revision; properties of the third sort should never, except in highly unusual circumstances, be included in an explanatory model. Concerning properties of the fourth sort, equidynamics has no special advice; such properties of course should and do frequently find their way into evolutionary models.

What gives equidynamic reasoning the power and flexibility to make judgments about probabilistic relevance across a vast range of enormously complicated systems? The answer has two parts. First, we humans have the ability to think about the proportional dynamics of the same vast range of systems, and in particular, to distinguish the kinds of factors that will make a difference to the proportion of trajectories leading to a given outcome from the kinds that, though they may shuffle trajectories around, make no difference to overall proportions. I have not sought to explain this ability; I simply take it for granted. Second, there is a strong relationship—made manifest in the microdynamic rule, the equilibrium rule, and so on—between proportions of trajectories and physical probabilities. The reasons for this relationship were sketched in part 2, and are investigated in greater depth in Strevens (2003, 2011). It is the second of these parts that accounts for our ability to untangle and find our way through complexity, and the first that accounts for the range of our equidynamic reasoning—its applicability to dice, deer, dodos, and dioecy.

11

INACCURACY, ERROR, AND
OTHER FLUCTUATIONS

Writing about the Battle of the Somme in his widely admired book *The Face of Battle,* John Keegan reasons as follows about the patterns of wounding from enemy fire:

> The chest and the abdomen form about fifty per cent of the surface of the body presented, when upright, to enemy projectiles; skin surface covering the spine and great vessels—the heart and the major arteries—is less than half of that. We may therefore conclude that about a quarter of wounds received which were not immediately fatal were to the chest and abdomen. (Keegan 1976, 273)

Why should the frequency with which a part of the body is struck by bullets be proportional to its area? Keegan does not say; he takes it for granted that the equation of frequency with area is a reasonable rule of thumb.

His thinking, and that of the readers who accept its prima facie validity, is surely based on implicit equidynamic reasoning. The inference in question has two steps. First, the initial conditions determining the spread of bullets from the rifles, machine guns, and so on at the Somme are assumed to be distributed microequiprobabilistically. Since the weapons are operated by their targets' fellow humans, this follows directly from the rule of physiological microequiprobability, which defeasibly warrants the assumption that the product of any physical human action is so distributed. (The rule was briefly mentioned in section 9.4, and will be treated fully in section 12.2.)

Second, under broadly Somme-like conditions, the range of initial conditions corresponding to impact points spanning the width of a human body—the distance between conditions resulting in a hit to one side of the body and a hit to the other—will be "microsized." For example, at a range of 100 meters, a rifle pointing at the extreme left of a typical torso would need to be rotated through an angle of about a quarter of a degree to point at the extreme right of the same torso. Holding the back of a rifle steady, this rotation would be accomplished by moving the front about 4 millimeters to the right. Supposing that this distance is physiologically "micro" (again, under Somme conditions), the probability distribution over the entire range of initial conditions resulting in a hit to a particular target will be approximately uniform. Thus the probability with which any area on the body is struck will be roughly proportional to its area.

The inference is accomplished, then, by invoking the physiological microequiprobability rule and doing some simple proportional dynamics, the latter hinging on shooting's sensitivity to initial conditions. Across the sciences, I hope to persuade you in this chapter, equidynamics is called upon to forecast the distribution of, and so to give science the tools to withstand, error, inaccuracy, fluctuation, and all other manner of minor spoiling variation.

11.1 Lining Up the Stars

"Suppose the rifle replaced by a telescope duly mounted," wrote John Herschel (1850, 18), comparing an astronomer's observational errors, such as the distance between the recorded position of a star and its actual position, with the aberrations of a sharpshooter. Herschel went on to provide his remarkable a priori derivation of the Gaussian distribution of measurement error, the same derivation that inspired, so it seems, Maxwell's "official" derivation of the molecular velocity distribution.

That was in 1850. My story begins hundreds of years earlier, when astronomers made their first attempts to grapple in a principled way with the problem of observational error—attempts from which the entirety of modern statistical thought grew.[1]

Point a telescope at a star and measure its position at the same time on the same day for several years running and you will get several slightly different results, largely due to inaccuracies in the process of measurement. What to do with these results? Take the measurement made under the best conditions, says one school of thought: the clearest sky, the calmest weather, the steadiest hand, the most serene mind. Not so, says the statistical school: average the measurements and you will have a value that is better than any single actual measurement, because "random errors tend to cancel one another" (Stigler 1986, 28).

In 1750, the technique of taking the average—usually in the sense of the arithmetic mean—of several observations made under similar circumstances as the best estimate of the observed quantity's true value was well established. (Plackett [1958] tracks it back to Tycho Brahe, and finds precedents in Hellenistic astronomy.) The method's appeal rests at least in part, as Stigler observes, on the maxim of statisticalism, that errors tend to cancel out. Why think that they cancel? Among early averagers, Galileo appeals to the principles that observations are "equally prone to err in one direction and the other" and if made carefully are "more likely to err little than much" (Galilei 1962, 290–291). Add the characteristically unspoken assumption of the independence of errors, and you have the premises you need to justify the expectation of canceling out, and so the method of setting your estimate equal to some sort of average.[2] Remember them.[3]

"Take the average," or more specifically "take the mean," is a useful method of estimation only if you are able to observe directly the quantity you wish to measure. Many astronomical measurements, however, are indirect: they seek to quantify the parameters of a planet's orbit, for example, or the orientation of the moon relative to the earth. In such a case, you face a problem of the following sort. You know (from physical first principles) that the value α that you wish to measure is related to observable quantities x and y by a certain mathematical formula, say $y = \alpha \sin x - \alpha \sin \theta \cos x + \beta$, where β and θ are also unknown.[4] Observation gives you various pairs of values of x and y; you must then use these values along with the equation to estimate the value of α (and collaterally, of β and θ).

Suppose, for example, that you make three observations of x and y, yielding pairs of values (x_1,y_1), (x_2,y_2), and (x_3,y_3). You can then infer values for α, β, and θ by solving the simultaneous equations

$$y_1 = \alpha \sin x_1 - \alpha \sin \theta \cos x_1 + \beta$$
$$y_2 = \alpha \sin x_2 - \alpha \sin \theta \cos x_2 + \beta$$
$$y_3 = \alpha \sin x_3 - \alpha \sin \theta \cos x_3 + \beta$$

where, note, α, β, and θ are the unknowns and the x_is and y_is are the knowns, having been determined by direct observation.

Three observations are enough to solve the equations, yielding determinate values for α, β, and θ. But what if you have made twelve observations, and so have twelve pairs of values and thus twelve equations? Solve any three of these equations and you get values for your parameters, but because of measurement error, you will get different values depending on which three you choose.

How to proceed? You might select the three pairs of measurements in which you have the most confidence. But the "statistical mind" will prefer to take advantage of all the empirical data, by finding some kind of average value for α, β, and θ. The analogy with the mean suggests the following procedure: solve every possible set of three equations to determine values for the three parameters, then take the average of these values. In practice, this is not feasible for equations of any complexity: given twelve equations with three unknowns, you would have to solve 440 sets of simultaneous equations. In 1750, while attempting to determine the libration of the moon (the wobbling of the face that it presents to the earth), Tobias Mayer took a more practical approach: he divided his equations into as many groups as he had variables to solve for, summed the equations in each group, and solved the resulting equations (Stigler 1986, chap. 1).[5] In the case at hand, you would divide your twelve equations into three groups of four and then add the four in each group together, getting from the first four observations, for example, the equation

$$\sum_{i=1}^{4} y_i = \alpha \sum_{i=1}^{4} \sin x_i + \alpha \sin\theta \sum_{i=1}^{4} \cos x_i + 4\beta .$$

That leaves you with just three equations, the sums of each of the three groups, determining unique values for α, β, and θ that take into account all twelve observations. In 1788, Laplace did much the same thing. (Stigler [1986, 37] explains the advantages of Laplace's procedure.)

Mayer and Laplace gave no formal justification for their procedures, but as in the case of taking the mean, two considerations seem to have been foremost: first, that the errors in observation are symmetrically distributed around the true value, and second, that the errors are independent. This much seems to underlie Mayer's claim that, the more observations of x and y are made, the closer his technique comes to yielding the true value of α.

The more modern way to think about the problem is as one of curve-fitting: find the values of α, β, and θ such that the curve $y = \alpha \sin x - \alpha \sin \theta \cos x + \beta$ "best fits" the observations; these values are then your

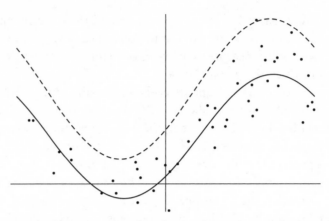

Figure 11.1: Finding the curve that best fits the observed pairs of values: the solid curve is a better fit than the dashed curve

best estimate of the true values. In graphical terms, plot your observations as points on a graph of *x* versus *y*; you want to find the values of α, β, and θ defining a curve that comes as close to the points as possible (figure 11.1).

Roger Boscovich, attempting in 1760 to determine the oblateness of the earth, was the first to take this approach: he prescribed, as a desideratum of an estimate of a set of unknown parameters, that they determine the curve that minimizes the sum of the distances between the curve and the observed data, or in statistical jargon, that minimizes the sum of the "residuals" (Stigler 1986, 46). Laplace later used a similar technique to deal with the same problem, and then in 1805, Adrien-Marie Legendre suggested minimizing the sum of the squares of the residuals—the "method of least squares."[6] The technique was adopted across Europe in short order: by 1815 it was "a standard tool in astronomy and geodesy in France, Italy, and Prussia; by 1825 the same was true in England" (Stigler 1986, 15).

The appeal of Legendre's method seems not to have been diminished by his providing only the barest attempt at justification:

> We see, therefore, that the method of least squares reveals, in a manner of speaking, the center around which the results of observations arrange themselves, so that the deviations from that center are as small as possible (Legendre 1805, 75; Stigler's translation).[7]

But other writers had been concerned for some time with the need to provide an argument for the use of averaging procedures, ideally an argument that singled out one such procedure as superior to every other.

The first such attempt was published by Thomas Simpson in 1755. It was the simplest of estimation procedures—take the mean as the best estimate of several observations of the same quantity—that Simpson wished to justify, by showing that the average of several observations is likely to be closer to the true value than a randomly selected single observation (Stigler 1986, 90).

To prove such a result, Simpson assumed a specific physical probability distribution for observation errors, that is, a specific "error curve," namely the triangular distribution shown in figure 11.2.[8] He used the distribution to calculate the likely divergence between the mean and the true value for a given number of observations—hence, the likely error in an estimate provided by "take the mean"—and he compared it to the likely error in a single observation, attempting to show that "take the mean" has the smaller expected error, and even better, that its expected error decreases as the number of observations increases (though as Stigler remarks, Simpson's mathematics falls short of this aspiration).

Shortly after Simpson's argument was presented at the Royal Society, Thomas Bayes formulated a pertinent criticism.[9] Simpson assumes that the distribution of errors will be symmetrical, writes Bayes, but

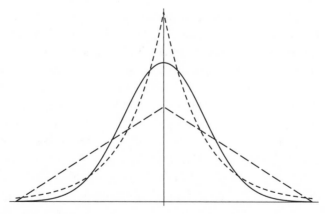

Figure 11.2: Three possible error curves centered around zero error: Simpson's triangular distribution (long dashed); Laplace's distribution (short dashed); Gauss's distribution (solid)

this assumption may well not hold if your measuring instrument has some specific imperfection. In response, Simpson revised his paper to make the grounds of the symmetry assumption explicit. His demonstration supposes, he writes,

> That there is nothing in the construction, or position of the instrument whereby the errors are constantly made to tend the same way, but that the respective chances for their happening in excess, and in defect, are either accurately, or nearly, the same (Stigler 1986, 95).

Both Bayes and Simpson traveled effortlessly from physical symmetries in the observation setup to symmetries in the distribution of errors.

Laplace tried a more sophisticated approach in 1774 (Laplace 1986). He aimed to find the estimation method that would minimize the expected error—that is, that would provide an estimate whose expected difference from the true value would be minimal. To make this calculation he needed, like Simpson, a hypothesis as to the form of the error curve. He took it for granted that the curve would be symmetrical, and then used further considerations of symmetry (that the rate of change of the curve should be directly proportional to the rate of change of its derivative) to infer what is now sometimes called the Laplace distribution: $\phi(x) = \frac{c}{2} e^{-c|x|}$ (figure 11.2).

It was Gauss (1809) who introduced the error curve as we know it today, the Gaussian or normal distribution $\phi(x) = \frac{c}{\sqrt{\pi}} e^{-c^2 x^2}$ (also shown in figure 11.2).[10] He too claimed that the error curve could safely be assumed to be symmetrical. His derivation of the normal distribution did not depend on this premise, however, but took a less secure and much criticized route: he assumed that the error curve is such that the probability of obtaining any particular set of observations is maximized by the hypothesis that the true value is equal to the observations' mean. He assumed, that is, that the error curve is such that the hypothesis with the highest likelihood on the evidence is always the hypothesis that the true value is equal to the mean. (He also assumed the stochastic independence of the errors; that is, he assumed that the probability of any set of observational errors is the product of the probabilities of the individual errors.)

Gauss went on to show that a necessary condition for the truth of this very strong assumption is that the errors have a Gaussian distribution (centered on zero), and further, that if the errors did have such a distribution, the method of least squares (which, when a single value rather than the form of a curve is to be inferred, amounts to "take the mean") would be guaranteed to provide the estimate that maximized the probability of the observed errors. The Gaussian error curve and the method of least squares were thereby co-enshrined as the chief dogmas of the nineteenth-century statistics of astronomical measurement.

After reading Gauss's derivation, Laplace realized that his own recent work on probability might supply a far better justification of the Gaussian error curve than Gauss himself had offered. The discovery in question was Laplace's celebrated central limit theorem (at that point still unpublished), which showed that the distribution of the sum of many independent random variables will, under a wide range of conditions, approximate the Gaussian distribution.

From the theorem Laplace concludes that, whatever the probability distribution over individual errors, the probability distribution over the mean observation error will be approximately Gaussian provided that there are many observations, the error in each is independent, and "in each observation, positive and negative errors are equally likely"; that is, the error distribution is symmetrical around zero (Laplace 1810). Thinking that this alone is not enough to justify the method of least squares—because a vindication broadly based on Gauss's requires a Gaussian distribution over individual errors, not merely over their mean—he goes on to provide the following somewhat tortuous vindication of the method. Suppose, he writes, that your observations can be divided into many sets, each subject to its own law of errors—presumably because all observations in a given set are produced using roughly the same instruments under roughly the same circumstances. The mean of each set has, by the central limit theorem, a Gaussian distribution centered on the truth, if the assumptions above apply—that is, if the sets are large and the errors are independent and symmetrically distributed. Laplace then proposes applying a variant of the method of least squares to these means, as though they were themselves individual observations. (In this variant, the means are first adjusted so as to compensate for the differences between the different Gaussians.)

This line of thought was taken further and at the same time simplified, though at the cost of stronger empirical assumptions, by Gotthilf Hagen, whose "hypothesis of elementary errors" posits that errors in individual observations are composites of many smaller, independent, symmetrically distributed perturbations or fluctuations, and concludes that the error distribution for individual observations is itself Gaussian (Hagen 1837).[11]

I will bring the chronicles to a close there, having told more than enough of the story to raise an important question about the history of statistics that equidynamics is in an excellent position to answer.

As Stigler (1986, 110) observes concerning the treatment of the error curve in astronomy,

> All early workers, from Simpson on, accepted it as given that the curve should be symmetrical and that the chance of an error should decrease toward zero as the magnitude of the error increased.

(The same is true for Galileo, writing in the previous century.) I have added that the errors' stochastic independence was also universally supposed. The preceding historical sketch shows how important these assumptions were in the development of statistical reasoning. Every thinker relies on symmetry and independence, whether to inspire confidence in their otherwise unjustified methods of estimation (take the mean, aggregate equations, least squares) or to derive a justification of the methods from the error curve. The sole exception is Gauss, who did not rely on symmetry but whose derivation was, for all its impact, widely thought to be question-begging. Without the security of the symmetry and independence assumptions, it is hard to see how the development of statistical thinking would have been possible.

So what is the source of the suppositions of symmetry and independence? Equidynamic thinking, I propose. Let me explain how, focusing on the symmetry assumption; I will have something to say about independence later.

Consider first two nonequidynamic explanations of the symmetry assumption's persistence through the early history of the statistics of measurement. According to the first explanation, the assumption is based on the epistemic principle of indifference rather than on equidynamic reasoning: since there is no known direction of bias, thinkers reason, a symmetric epistemic probability distribution over error is warranted.

While this train of thought may have had some influence, what explicit discussion we have of the symmetry assumption fits it badly. Bayes objects to the symmetry assumption on the grounds that an imperfect measuring instrument may give rise to an asymmetric distribution of errors. There is nothing epistemic about this worry; the problem is that asymmetry in the instrument's physical setup, whether or not it is known to the reasoner, will cause an asymmetry in the actual frequencies of errors. Further, Laplace is explicitly concerned with finding the *correct* error curve, by which he clearly means not correct in its reflecting his ignorance, but correct in its reflecting the actual distribution of errors, or facts about the causes of this distribution. (For Laplace's "objectivist" side, see section 3.2 and Daston [1988, 218ff].)

Much later, Herschel justifies the symmetry assumption on the basis of ignorance, in the derivation of the Gaussian error curve that inspired Maxwell. But as noted in section 2.2, he slips almost immediately into a nonepistemic, purely physical mode of reasoning: "supposing no bias, or any cause of error acting preferably in one direction, to exist . . .". Herschel's talk of ignorance is, I suggest, ideology; the slip exposes the equidynamic means of inferential production that underlies even his thought.

The second nonequidynamic explanation of the symmetry assumption goes to the other extreme. On this view, the symmetry of the error distribution is founded on statistical knowledge: symmetry is assumed because it is known that errors are in fact typically distributed symmetrically.

This is the correct explanation, I think, of another assumption about the distribution of errors, namely, that the larger the magnitude of an error, the smaller its probability, which is explicitly supposed by Galileo, Laplace, and others (Laplace 1986, 370). Although equidynamic considerations may play some role in supporting the assumption (more on this below), it is surely based also, and at least equally as much, on everyday experience of measurement error.

With respect to the assumption of symmetry, by contrast, observed frequencies have a lesser role to play. It is not immediately discernible (or even, as you will see shortly, universally true) that errors of a given magnitude equally often occur in any direction, and there appear to have been very few attempts to compile statistics concerning observational error in astronomy. F. W. Bessel published some data supporting the Gaussian hypothesis about the distribution of errors in 1818 (Stigler 1986, 202–204), but by that time the symmetry assumption had been in play for seventy years; its most important contributions to the study of error were behind it. As Stigler remarks, the tedious hard work needed to collect the numbers seemed justifiable only once the power of statistical methods was established.

I conclude, then, that the symmetry assumption was based neither on ignorance of the direction of measurement bias nor on the statistics of actual measurement, but on something empirical but nonstatistical, namely, partial knowledge of measurement's physics.

What information did eighteenth- and nineteenth-century astronomers have about the physical processes involved in observation? As you might expect, they had a keen awareness of such matters. A useful source is Bessel's (1838, section 10) enumeration of thirteen (somewhat overlapping) causes of error in measuring the position of a fixed star.

Bessel's causes may be divided into three classes. First, there is noise that buffets the rays of light before they reach the measuring instrument, such as the "twinkling" of a star caused by atmospheric fluctuations.

Second, there is variability, noise, or error caused by imperfections in the measuring instrument itself, as when for example there are optical flaws in a telescope's mirror. Such imperfections might be caused either by imprecise construction or by environmental effects such as uneven warming of the apparatus (causing some parts to expand more than others).

Third, there are errors introduced by the observers themselves. For example, to determine exactly where along a marked scale a star's position falls, an astronomer may use a vernier, a movable scale marked with finer divisions than those on the main scale. The vernier should be lined up exactly with the star, but due to the imprecision in human coordination of hand and eye there is typically some slight misalignment.

In many cases, the observer's contribution to variability in measurement comes not so much from the imperfection of the human machine as from arbitrary choices that the observer must make because of variability introduced earlier in the measurement process. There is no exact fact of the matter, for example, as to when a twinkling star is lined up with a reticle, since the apparent center of the star will move around with the fluctuations that cause the twinkling; the observer must therefore simply deem a certain alignment to be final.

How are such errors likely to be distributed? In most cases, there are good equidynamic reasons to infer symmetry, if not with complete confidence.

Consider environmental noise, for example. We would now understand both atmospheric and thermodynamic fluctuations much as Maxwell did, that is, as the product of interactions between particles in disordered motion, and so would apply the same equidynamic reasoning to inferring the distribution of such fluctuations as young children do in Téglás et al.'s bouncing balls tasks. Eighteenth-century scientists did not think about air and heat in this way, however; the caloric theory of heat was widely accepted, and the kinetic theory of gases was a minority view.

Suppose, then, that air is a kind of substance, we know not what. That there are atmospheric fluctuations shows that the substance is, on a small scale, moving in a disordered way. Presumably, then, the substance can be divided into small "packets" (which may or may not be individual particles), each moving somewhat independently of the others. We can think about these packets in much the same way that we think about balls in an urn or gas molecules: given some weak assumptions about the interactions between the packets, we can apply the microdynamic rule to infer the microequiprobability of packet position, and then apply the equilibrium and uniformity rules to infer that a packet moving at a given speed is equally likely to be moving in any direction (sections 7.1, 7.2, and 7.3).[12] The same thinking might be applied to fluctuations in caloric fluid, the stuff of heat.

What about the human factor? Suppose you are lining up a vernier with a star. How do you proceed? You move the vernier in the right direction by about the right amount. If the correspondence to the star's position is close enough (more or less indiscernible to the unaided eye, let's say), you are done. If not, you repeat the process: move the vernier in the right direction and ask yourself whether it is close enough. Take the final such movement, that is, the movement that shifts the vernier to the position that you judge adequate for the purposes of measurement. There will be some variation in the end point of this final movement. Sometimes the vernier will be slightly too far to one side; sometimes, it will be slightly too far to the other. This variation is, in physiological terms, on the "micro" scale; thus, by the physiological microequiprobability rule, it is reasonable to suppose that it is uniformly distributed, hence symmetrical around the correct position (assuming that, in discerning whether the vernier is sufficiently well aligned, you tolerate equal amounts of deviation on either side of the correct position).

By much the same reasoning, you can infer that the joint distribution over any two vernier positioning errors is uniform, which is sufficient for and so justifies the assumption of independence. It is quite possible, however, that early statisticians did not reason in this way, but rather assumed the stochastic independence of separate errors on the grounds of the surmised causal independence of error-creating processes, using what I call in section 13.2 the independence rule.

Judgments of symmetry and independence in the errors are, I should conclude by saying, corrigible: the equidynamic judgments rehearsed above depend on assumptions about the physics of measurement that may be false. Indeed, in some cases they are false. The astronomer William Rogers noted in 1869 that his recordings of the time that a given star passed a designated point—corresponding to its crossing a meridian line in a fixed telescope—were faster when he was hungry; more generally, some astronomers tended to record earlier times than others (Stigler 1986, 240–241). This does not, of course, undercut the validity of the symmetry assumption in the range of situations where its physical premises turn out to be correct. In any case, my aim here is not so much to vindicate as to explain the thinking of the early statisticians; what matters is that their assumptions of symmetry and independence, without which the development of statistics would have been very different and I think greatly retarded, hinged on equidynamic reasoning.

11.2 Atmospherics

This planet's weather is caused by the complex structure of its atmosphere's lowest level, the troposphere. The circulation of the troposphere within each hemisphere, north and south, occurs in three zones: from the equator to roughly the 30th parallel (the line of 30 degrees latitude) is the Hadley convection cell, responsible for the trade winds blowing from the northeast and southeast (in the Northern and Southern Hemispheres, respectively); from roughly the 60th parallel to the pole is the polar convection cell, also responsible for easterly surface winds (though weaker than the trades); and between the 30th and 60th parallels is the Ferrel cell, driven by and rotating in the opposite direction to the Hadley and polar cells, thus creating surface winds that are on the whole westerly.

Between the cells, fast and narrow westerly currents of air form near the top of the troposphere—the jet streams. The polar jet, which is the stronger of the two, lies at the border of the polar and the Ferrel cells, while the subtropical jet lies where the Ferrel and Hadley cells meet. The polar jet often takes a meandering shape, corresponding to the meandering border between the polar and Ferrel cells; there is a degree of irregularity in the border between the Ferrel and Hadley cells as well.

Of the three cells, the Ferrel has the most complicated structure, and so the mid-latitudes have the most complex weather. Most notable are the formation, at the border of the polar and Ferrel cells, of fronts and accompanying cyclones ("lows").

By the middle of the twentieth century, these structural facts about the atmosphere were well established, thanks in part to the use of weather balloons and to the observations of military aviators. The formation of the Hadley cells had been posited to explain the trade winds long before by Edmond Halley and then (in 1735, incorporating the Coriolis effect) by George Hadley. But the other elements of tropospheric dynamics were poorly understood: possible mechanisms for the formation of cells, jets, fronts, and cyclones were formulated, but it was not known which mechanisms were correct or why.[13]

Many meteorologists were convinced that the structure of the troposphere could be explained by the basic principles of fluid mechanics, supplemented by other physical principles governing processes such as the radiation of heat into space. But the relevant system of differential equations, even in its simplest form, would not yield to analysis. The best that could be expected was a numerical solution, which would calculate the effect of the equations, or rather an approximation thereof, for a particular set of initial and boundary conditions.

The difficulty of this computational task is well illustrated by Lewis Fry Richardson's attempt, in 1917 while driving an ambulance on the Western Front, to retrodict six hours of weather patterns observed by a pan-European release of weather balloons in 1910. The calculations took "the best part of six weeks" (Richardson 1922, 219), and resulted in an incorrect prediction for deep reasons explained by Lynch (2006). Richardson suggested wryly that future forecasts might be undertaken by massed ranks of humanity arranged in the form of the surface of the globe, crunching numbers in unison under the direction of a leader acting as "the conductor of an orchestra in which the instruments are slide-rules and calculating machines" (Richardson 1922, 219)—a better use of manpower than trench warfare, at least.

Then the digital computer arrived, and for human calculators it was possible to substitute vacuum tubes, then transistors, then integrated circuits. The first mechanized weather simulations were performed by punch card machines; soon enough, thanks in part to the influence of John von Neumann, the ENIAC computer was recruited to the cause, to be superseded in due course by more sophisticated machines.[14]

In 1955 Norman Phillips constructed the first general circulation model of the earth's atmosphere—a model covering a wide enough swathe of territory, and with a long enough time frame, to capture the structural features enumerated above: the Hadley, polar, and Ferrel convection cells; the jet streams; the formation of fronts and cyclones; and so on (Phillips 1956). Phillips' idea was to start the model with a static atmosphere having none of these features; to introduce solar heating, warming the air in the tropics and so creating the conditions that enable convection; and then to see what sort of atmospheric structure developed (with the help of a "random disturbance" introduced partway through the simulation). The results were striking: the simulated atmosphere indeed had the three-cell structure, with jets, fronts and cyclones. Further, analysis of the simulation cast considerable light on the strengths and weaknesses of the various theories previously offered to explain these structures.

Yet all of this was worse than useless unless Phillips' model was relevantly similar—similar in its structure-generating properties—to the real atmosphere. On that count, problems: the model's dynamics, writes Lewis (2000, 119), "exhibited an almost irreverent disregard for the complexities of the real atmosphere." It self-consciously left out, among other things, the effects of "small-scale turbulence and convection, the release of latent heat, and the dependence of radiation on temperature, moisture, and cloud" (Phillips 1956, 157). Indeed, the model contained no representation whatsoever of such meteorologically important factors as mountain ranges and clouds, or even of the distinction between land and sea. Phillips did not have much choice. His computer, the IAS computer (so-called because it was developed at the Princeton Institute for Advanced Study), had only 1,024 words of random access memory and a further 2,048 words of magnetic storage, the latter used in large part to store the program. The entire atmosphere had to be represented in those 1,024 words of RAM, while leaving space for calculation. Phillips divided up the part of the atmosphere he modeled (notionally, a section of the Northern Hemisphere considered as a cylinder) into a 17×16 grid, with the air flow represented at just two levels for each sector of the grid. Consequently, the model used not many more than 500 variables.

You might think that, for all its heroism, such an effort could hardly be taken seriously: there was just too much missing from Phillips' model to regard the causes of its behavior as causes also of the behavior of the actual atmosphere. It is, you might think, the paradigm of a model that gets "the right results for the wrong reasons" (Edwards 2010, 154).

You would be mistaken. Phillips' model "provoked enormous excitement," inspiring a generation of climate modelers after Phillips (Lewis

2000, 118–119; Edwards 2010, 152). Clearly, then, many climate researchers regarded Phillips' model as likely enough to be relevantly similar to the earth's atmosphere that they were willing to take the bet that building comparable or slightly more sophisticated models would provide real meteorological and climatological knowledge. What was the source of their confidence?

What, I should first ask, is relevant similarity? Phillips' model is relevantly similar to the earth's atmosphere for the purpose of investigating a given structural feature—say, the formation of the three convection cells—just in case the elements omitted from or distorted by the model make no difference to whether or not the atmosphere has the feature in question. The very existence of the feature, then, and not merely some aspect of its behavior, must hinge on an element in order for that element to qualify as a difference-maker.

Difference-making is a technical notion related to explanatory relevance; Strevens (2008) gives an account of the difference-making relation, one of several in the literature, and discusses the question of what it means to say that a distorted factor makes no difference. No need to examine the intricacies of difference-making here, however. Raw intuition will be enough to pose my question: what reason did Phillips' readers have for thinking that the land-sea divide, clouds, or small-scale turbulence, made no difference to the existence of the tripartite cellular structure, or of the jet streams?

Let me ask about small-scale turbulence in particular. By narrowing the focus in this way, I will be putting aside almost all of the physical intuitions and other background cognitive machinations that go into deciding that a model is sufficiently similar to the world to command prima facie scientific respect, and following just one slender thread in the complex tangle of rationales required to make such a judgment—though in the hope, of course, that this single example might fulfill some of the promise of Blake's grain of sand.

Why think, then, that small-scale turbulence makes no difference to the existence of the aspects of the atmosphere's structure captured by Phillips' model? Begin with the converse question: under what conditions would small-scale turbulence undermine or displace, at least temporarily, large-scale atmospheric structure? I can see two. First, small-scale disturbances might individually be underminers of structure. Second, many such disturbances, coming together in a coordinated way, might add up to a large-scale undermining disturbance.

It was clear to Phillips' contemporaries that the immensely powerful forces driving the formation of large-scale structures such as convection cells would shrug off small disturbances.[15] As for large disturbances, their effect of course depends on their scale, timing, and location. But I think it is safe to say that 1950s meteorologists thought, as their successors do today, that large aggregate disturbances can be ignored because there are none—or, more exactly, that the chances of many small-scale disturbances adding up to something large enough to disrupt the gross structure of the troposphere are, because of small-scale distributions' generally scattered and uncoordinated nature, negligible. The record of criticisms of Phillips' model certainly shows no concern about eventualities of this sort (Lewis 2000, section 4).

The same assumption is even more crucial when modern general circulation models are used to forecast the near future's weather. A disturbance sufficiently great to prevent the formation of the jet stream would require a miraculous alignment of small-scale disturbances (not to mention the formation of compensatory movements elsewhere to ensure the conservation of the atmosphere's angular momentum). A disturbance large enough to throw off a short-term forecast is much easier to come by. But still forecasters suppose that the formation of such disturbances from the coincidental coordination of small-scale disturbances is extremely unlikely. In this assumption, they appear to be correct: the challenge of short-term weather forecasting lies more in representing existing medium-scale effects than in the possibility that other effects on the same or greater scale may emerge without warning from nowhere. Thus, as the representational challenge has been met, the accuracy of weather forecasts has improved enormously, though even the finest-grained contemporary models represent the atmosphere at scales of tens of kilometers, and so in effect ignore small-scale disturbances completely.

Forecasters' confidence in the vanishingly small probability of freak disturbances has been empirically vindicated, then, but this statistical assurance is a late arrival. Our confidence in the low likelihood of such disturbances is therefore not based on the frequencies. (Indeed, the occurrence of sudden, violent, inexplicable weather events looms large in folk memory.)

Where does the confidence, in that case, come from? The matter seems almost trivial: what are the chances of thousands of episodes of small-scale turbulence all pulling the same way at the same time? Almost zero, obviously. Obvious indeed, but why? Show your working.

A little reflection reveals that, while the assumption of negligibility is psychologically compelling, there is no easy way to establish its truth. Indeed, there is no easy way to establish anything much about as complex a system as the earth's atmosphere. When you have an "obvious" assumption that cannot be easily and instantly justified, something is going on below thought's conscious layer. That something—the inference that gives us the confidence to ask rhetorically, of a freak disturbance, "What are the chances of that?"—is, I propose, equidynamic. From the physics of the situation, we can use the principles of equidynamics to infer certain facts about the probability distribution of small-scale disturbances, facts that make the large-scale coordination of such disturbances extremely unlikely.

On what facts is the equidynamic inference based? Small atmospheric disturbances may in principle be described using the probabilistic apparatus of kinetic theory; can a low probability for freak disturbances be derived from Maxwellian considerations, then? Not directly: the individual causes or components of a freak disturbance—isolated thunderstorms or changes in ground reflectivity due to vast roving herds of animals, for example—may be small on a meteorological scale, but they are still from a statistical mechanical perspective macrolevel phenomena, and so do not derive a probability distribution from kinetic theory. Can the microequiprobability rule, which defeasibly warrants an assumption of microequiprobabilistic distribution for any variable (sections 9.4 and 12.1), be put to use here? No; unassisted, the rule cannot deliver judgments about the low probability of freak disturbances, for much the same reason that statistical mechanics cannot: the scale of the small disturbances is not "micro" enough for the rule to make direct pronouncements about their distribution.[16]

What is needed to complete the story is, I think, some understanding of the physics, and more especially the physical symmetries, of weather creation. Small-scale turbulence is generated—so meteorologists may reasonably suppose—by the amplifying effects of the initial condition–sensitive dynamics of fluid mechanics (evident long before "chaos theory" acquired its name) or by the confluence of smaller-scale instances of turbulence determined in turn in the same way. The microequiprobability rule supplies a probability distribution over the perturbations that are the seeds of such disturbances, and thus over the disturbances themselves. The form of the distribution will typically depend on the contours of the immediate environment, that is, on such matters as the shape of the local landforms and the velocity of local prevailing winds. There is no single probability distribution over small-scale disturbances, then; rather, there is a probability distribution for every set of local circumstances. Local circumstances change—with the tides, with the seasons, with caribou migrations, with the visitations of El Niño and La Niña. The totality of probability distributions determining the overall distribution of small-scale disturbances is therefore always on the move.

Very little need be known about these distributions, however, to derive a low probability for freak large-scale disturbances. Two properties of the distributions are, I think, sufficient for the derivation: first, that the distributions for the most part allow a fairly broad range of directions and intensities, and second, that they are approximately independent. From there, it is easy to infer that the probability of the coincidental alignment of a large number of small-scale disturbances is negligible. Hence, "freak." What is the source of the probabilistic assumptions? The microequiprobability rule and proportional dynamics, in concert with some seat-of-the-pants but nevertheless empirically plausible small-scale atmospheric dynamics.

A few remarks. First, it need not be assumed that every occasion for a small-scale disturbance is capable of creating "a fairly broad range of directions and intensities." Some disturbances reliably take on a fixed direction, for example, winds generated by geographical features such as valley exit jets. Geographical considerations alone tell against the large-scale coordination of such phenomena. Second, my suggested derivation of a negligibly low probability for freak disturbances is rather vague. This is as it should be: every meteorologist likely reasons about these matters in a somewhat different way (insofar as their reasoning is conscious at all). What I have provided is a template. The probabilistic premises on which the low probability is based are so weak that they may be reasonably reached by many moderately different epistemic routes. Third, the reasoning is, as is usual with equidynamics, provisional.

11.3 One over Many

As for the weather, so for macromodeling in general. Any macrolevel process is in principle vulnerable to disruption by a spontaneous alignment of large numbers of microsized variations. If your model for such a process keeps track only of macrolevel quantities, you are implicitly supposing, by applying the model to the real world, that such spontaneous alignments have a negligible probability (according to the standards of negligibility appropriate to your application). You are supposing, in other words, that such alignments are freak events.[17]

The negligibility assumption is typically quietly made. But it is crucial, a "little transmission gear turning noiselessly that is one of the most essential parts of the machine."[18] A method whose function is to make large obstacles invisible is unlikely to receive its due share of the credit. Do not perpetuate this injustice; epistemically responsible modeling could not proceed without the negligibility assumption or without the equidynamic principles on which it is based.

Consider the models of population ecology describing the changes in the numbers of predator and prey species sharing the same habitat. As formulated by Lotka (1956) and Volterra (1926), these models take the form of pairs of differential equations, entirely deterministic. Specify the precise number of foxes and rabbits, then, and the Lotka-Volterra models will purport to tell you precisely how many foxes and rabbits there will be in a year's time, just as Phillips' model purports to tell you, for some initial atmospheric state, what the atmosphere will be doing in a week or a month's time.

Like Phillips' model, Lotka-Volterra models are used more for qualitative than quantitative purposes—to explain why certain kinds of ecosystems are stable under small perturbations, for example. Even uses such as these presume, however, that the models make short-term predictions that are at least approximately correct, most of the time. Thus they presume that large fluctuations in population arising from causes not represented in the model are rare. For this reason, the models are considered inapplicable to systems that experience regular external shocks (unless of course the shocks are built into the model); likewise, they should be considered inapplicable if a system's populations are subject to large spontaneous fluctuations, including those that might result from the coincidental alignment of the many small fluctuations in population due to internal features of an ecosystem not represented by the Lotka-Volterra apparatus.

Are population ecologists aware of such small fluctuations? Are they therefore aware of the in-principle possibility of large fluctuations? Do they think of the small fluctuations stochastically? From this stochastic thinking, do they conclude that large spontaneous fluctuations are overwhelmingly improbable, and so may be safely dismissed as potential spoilers for Lotka-Volterra models' validity?

There is direct evidence for affirmative answers to the first three questions. Lotka, when introducing the mathematics of predator-prey population change, suggested that the relation between the dynamics of populations and the dynamics of individual organisms (their movement around their habitat, their feeding, and so on) should be modeled on the relation between the macrolevel and the microlevel in statistical mechanics (Lotka 1956, 50). Later in the same work he developed a rudimentary probabilistic microdynamics, deriving the rate with which predators capture prey from some simple probabilistic assumptions about the paths taken by organisms through their habitats (pp. 358–361). It is a small and unremarkable step from these assumptions to the conclusion that spontaneous fluctuations in population—fluctuations not represented in the Lotka-Volterra dynamics—are ubiquitous but almost never large enough to undermine the validity of the models.

Likewise, I showed in chapter 9 that nineteenth-century biologists were both acutely aware of the "chance" aspects of survival and reproduction in nature's melee, and quite willing to conceive of these aspects in terms of physical probabilities based in equidynamics. It seems eminently plausible, then, that population ecologists, like meteorologists, assume implicitly that large spontaneous fluctuations pose no problems for their models because of their physical improbability, the negligible size of such probabilities being equidynamically derived.

And so on, for any discipline or subdiscipline that studies macrolevel variation in a system subject to microlevel perturbation. Such systems may be more like the climate than like a predator-prey ecosystem in their stochastic profile. They might not be susceptible to the kind of generalized statistical mechanics—what Strevens (2003) calls enion probability analysis—that intrigued Lotka, then; but still equidynamics has a role to play in justifying the principled neglect of perturbation from below. Economics, for example, might seem the most self-consciously un-stochastic science in its conception of its microfoundations, but who seriously doubts that there is randomness or noise to be found in economic actors' behavior? There must be some reason other than wishful thinking to exclude such variation from economic models; the variation must be considered, rightly or wrongly, at all times minor and uncorrelated. Behind every great deterministic theory in the biological and social sciences is, I suggest, a stochastic—an equidynamic—rationale.

From the celestial sphere down to the Flanders mud, the strands of support and influence are subtle, but they reach in every direction. Paul

Samuelson, the greatest twentieth-century mathematicizer of economics, was strongly influenced by Lotka, observes Herbert Simon in a brief but captivating review of the posthumous reissue of Lotka's *Elements of Physical Biology*. In the 1930s, Simon reminisces, "the use of mathematics in social science was an oddity and the proponents of that use eccentrics" (Simon 1959, 493). These wandering souls and free spirits formed a sect, he continues, whose twin arcane texts and inspirations were Lotka's *Elements* and *Generalized Foreign Politics*—the latter a treatise by the pacifist ambulance-driver and weather-forecaster Lewis Fry Richardson, offering a quantitative macromodel of causes of conflict written in the hope that a mathematics of war might help to eliminate this greatest of all human errors.

IV

BEFORE AND AFTER

12

THE EXOGENOUS ZONE

Adult specimens of Poli's stellate barnacle *(Chthamalus stellatus)*—a common species of a genus mentioned several times in Darwin's *Origin*—vary in length from about 0.2 to 1.4 cm. Quickly: how will the probability of such a barnacle's growing to between 0.8 and 0.81 cm in length compare to the probability of its growing to between 0.81 and 0.82 cm? The two probabilities will surely not differ by much, you will reply. Your default (though defeasible) assumption is that the size of the organism is distributed microequiprobabilistically. *Natura non facit saltus.*

You make this assumption despite your not having the knowledge needed to derive microequiprobability from the biology of barnacle development: as far as equidynamics is concerned, the probability distribution over *C. stellatus* size is "epistemically exogenous," by which I mean that you know little or nothing about the way in which the distribution is generated. Nor, of course, do you have a close acquaintance with the relevant statistics.[1] To answer the question on the basis of a provisional supposition of microequiprobability seems reasonable all the same.

An assumption of microequiprobability for exogenous probability distributions has turned up previously several times over: in the treatment of stirring processes; in the justification of the microdynamic rule; in our reasoning about the biological advantage of freezing upon detecting a predator (though there it might be attributed to an application of the microdynamic rule). To unravel the assumption is this chapter's goal.

My questions are as follows. First, concerning the world: Are the initial conditions of microconstant and other processes typically distributed microequiprobabilistically, and if so, why? Are the distributions over some variables more likely to be microequiprobable than others? In what circumstances will a distribution's microequiprobability be undermined?

Second, concerning the inferential structure of equidynamics: Under what conditions do we assume microequiprobability for exogenous distributions? Does our confidence in microequiprobability vary with the nature of the variables in question? Do we recognize defeaters for our inferences to microequiprobability?

12.1 The Microequiprobability Rule

Microequiprobability is everywhere—so much so, that it is reasonable to assume that any physical quantity is microequiprobabilistically distributed unless there is some positive reason to think otherwise.[2] Or at least, the assumption of microequiprobability is reasonable with a few caveats and qualifications, to be spelled out in what follows.

There are two ways to assume microequiprobability: explicitly and implicitly. An explicit assumption adds to the reasoner's set of available premises the microequiprobability of a probability distribution. An implicit assumption manifests itself in an inference rather than a proposition: the reasoner puts to work, in some particular case, an inference method that is reliable only if some variable or other is distributed microequiprobabilistically. Let me focus in this section on the rules for making an explicit assumption; I will then ask to what degree the sophistications present in the explicit rules are present also in our implicit appeals to microequiprobability.

Defeasibly assume that any variable is microequiprobabilistically distributed: that would be the simplest rule for introducing explicit assumptions of microequiprobability. But the rule is unreasonable, because its reliability would require a logical impossibility. For every random variable that is microequiprobabilistically distributed, there is at least one other that is not—one other whose nonmicroequiprobability is entailed by the original variable's microequiprobability.[3]

Suppose, for example, that some variable v has a microequiprobabilistic distribution. Then the variable v^*, defined so that $v = v^* + (\cos \pi v^* - 1)/\pi$, is not microequiprobabilistically distributed, as figure 12.1 shows—Bertrand again.[4]

A statement of the ubiquity of microequiprobability must therefore restrict its scope in some way, as must a methodological rule licensing a defeasible assumption of microequiprobability. I favor a restriction, as elsewhere in equidynamics, to standard variables, which when they were first introduced in section 5.6 were claimed to show a tendency to dis-

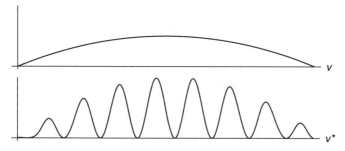

Figure 12.1: A microequiprobabilistic probability distribution over v (top) is equivalent to a nonmicroequiprobabilistic distribution over v^* (bottom)

tribute themselves microequiprobabilistically. (You will recall that the standard variables are roughly those variables that induce measures directly proportional to the SI units. Thus meters, seconds, and meters per second are standard variables, while gerrymandered versions of the same constructed as shown above are not.)

If microequiprobability is nearly ubiquitous among standard variables, then the following rule for introducing explicit assumptions of microequiprobability seems reasonable:

> Defeasibly assume that any physical probability distribution over a standard variable is microequiprobabilistic,

relative to a standard for "micro" on which a region counts as micro-sized if it spans a pretty small proportion (say, 2 percent or less) of the variable's full instantiated range. Call it the *microequiprobability rule.*[5]

It would be nice to have a less anthropocentric characterization of a standard variable. It would be nice to have a standard for "micro" that was less crude and perhaps sensitive in some ways to the variable in question. It would also be nice to have a reason for believing that micro-equiprobability is widespread, or even better, an explanation of standard variables' proclivity to get themselves distributed microequiprobabilistically. All of this will be forthcoming in section 12.3.

Observe that the microequiprobability rule as stated presupposes a physical probability distribution. Thus, it should be applied on a per-distribution basis. It does not have implications for the distribution of, say, length or distance or position generally, because there is no such thing as *the* distribution over position. It has implications only for cases in which there is a probability distribution over the length or position of some particular kind of object, or the distance between particular kinds of objects, in some particular context. It might, for example, license an assumption of microequiprobability for *C. stellatus* barnacle size or for the distance between gum trees in a certain eucalyptus forest. It does not license the assumption for the masses of protons, which do not fall under a physical probability distribution at all. Nor is the assumption licensed for the size distribution of the barnacle *Ibla cumingi,* first described by Darwin and also mentioned in the *Origin,* a radically sexually dimorphic species in which the females are (by volume) hundreds of times bigger than the males, which live as parasites upon them. Separate probability distributions apply, in this case, to the dimensions of the males and the females of the species; each may be assumed to be microequiprobabilistic, but relative to different standards of "micro."

Must you know that there is a probability distribution over a set of standard variable instantiations in order to apply the rule? No; fruitful and well-grounded reasoning can proceed in the absence of knowledge. General epistemology is not my topic here, but you might for example attach a certain subjective probability to a variable's having a physical probability distribution; the microequiprobability rule then licenses you to attach a similarly sized (though presumably slightly smaller) subjective probability to the variable's microequiprobability. Or if you assume defeasibly that you are dealing with a probability distribution, the microequiprobability rule licenses you to assume also that the distribution is microequiprobabilistic, subject to the same (and certain further) defeaters. (Advance warning: the condition that the microequiprobability rule applies only where a well-defined physical probability distribution exists will be weakened in section 12.5.)

According to the microequiprobability rule, it is reasonable to suppose defeasibly that an unknown physical probability distribution over a standard variable is microequiprobabilistic, provided that the distribution is epistemically exogenous—provided that you currently know little or nothing about the way in which the distribution is generated. The rule thus provides a way to get started in a state of incomplete probabilistic knowledge.

If you know something about a distribution's generation, your superior epistemic position imposes on you certain obligations. You ought not, for example, to assume a variable's microequiprobability if the process generating the variable has a dynamics that turns a microequiprobabilistic distribution over initial conditions into a nonmicroequiprobabilistic distribution over outcomes (as do, perhaps, some developmental processes in biology, for example those producing certain patterns of camouflage). More generally, if you know enough about the distribution's generating conditions to infer on independent grounds its microequiprobability or lack thereof, you know enough to render the microequiprobability rule's opinions irrelevant. This qualification and others like it may be subsumed under the microequiprobability rule's prefatory "defeasibly."

You also ought to proportion your confidence in microequiprobability to the evidence. If the variable assumed to be microequiprobable is known to be generated by a process that almost always produces microequiprobabilistic distributions—the case of human motor control, discussed in section 12.2, is one example—then you ought to be more sure of microequiprobability than in other circumstances.

I have so far been writing in the idiom of "ought," but my conclusions can be transferred, I think, to the domain of "is": real human reasoners follow the microequiprobability rule, respecting the strictures enumerated above.

So much for explicit assumptions of microequiprobability; what of implicit assumptions—that is, the use of methods that are reliable only when the relevant initial condition distributions are microequiprobabilistically distributed? Examples include the microdynamic rule and most likely the stirring rule for which I surmise the explicit premises are purely dynamic.

The principal question to ask, I think, is whether we make implicit assumptions with the same finesse as explicit assumptions. Do we abandon methods founded on implicit assumptions of microequiprobability, for example, when the initial conditions in question are generated by a process that transforms microequiprobabilistic conditions into nonmicroequiprobabilistic conditions, or by a process that is not stochastic at all?

In certain cases, the answer is surely yes. If we know that a coin is tossed with precisely the same loft and spin again and again, we can see that the outcome will tend to be the same every time, and so we will withhold our usual judgment about the probability of heads. It also seems possible that we—not only professors of probability but ordinary humans—exercise the same caution when statistics or other considerations suggest nonmicroequiprobabilistic variation in the initial conditions. But for this hypothesis I have no direct evidence.

The microequiprobability rule, I should emphasize, does not obviate the need for the rest of equidynamics. To yield conclusions of any great interest, it typically requires at least a dash of proportional dynamics, as in the majority rule. But even that is not enough to derive the probabilities of coin tosses and die rolls, let alone to think about molecular velocities or comparative fitness. For these latter feats of probabilistic inference we need all the equidynamic apparatus developed in previous chapters.

If the microequiprobability rule looks to make any part of equidynamics redundant, it is the microdynamic rule, since both rules license the same kind of conclusion—namely, the microequiprobabilistic distribution of a certain variable. But it is the microdynamic rule that defeats the microequiprobability rule, for the reason given above: the microequiprobability rule yields quietly in the face of facts about a probability distribution inferred from properties of its generating dynamics. In this case, where the two rules render the same verdict, it may barely be perceptible which gives way to the other; there is a practical difference between the two rules' conclusions, however: we can be more confident in microequiprobability inferred using the microdynamic rule than the same inferred using the microequiprobability rule.

Microequiprobability is everywhere—that is so far no more than a brute assumption, true perhaps, but without an explanation. The assumption seems eminently reasonable, of course, but that is because the microequiprobability rule is knitted into your brain. To investigate the causes of microequiprobability, then, is my aim in the next two sections, first on an intimate human scale and then with practically Platonic abstraction and grandeur.

12.2 The Human Touch

When equidynamics assigns physical probabilities to gambling devices—coin tosses, die rolls, urn draws, and so on—it assumes implicitly the microequiprobability of the relevant initial conditions, which are in these cases a suite of motions produced by human beings, the devices' "croupiers." Why do croupiers' tosses, shakings, rolls, and rummagings come in microequiprobabilistic form? The beginnings of an answer lie in research on human physiology.

Most such work examines goal-directed action—reaching for a glass, mousing to a menu, applying calipers to a barnacle, and so on. These actions display a predictable, statistical pattern of variation: when a movement pattern requires a certain degree of force, that force is produced with a standard deviation that increases linearly with its mean (Schmidt et al. 1979; Todorov 2002). Goal-directed muscular actions, then, are subject to a degree of "neuromotor noise" that affects their accuracy.

The same is true, I presume, of croupiers' actions, which are at least as variable as movements of the goal-directed kind. The variation is not entirely due to neuromotor noise—some variation may, for example, be introduced by conscious choice—but neuromotor noise will introduce a "smoothing" effect that guarantees microequiprobability.[6]

The source of motor noise is not yet known. One aspect of the noise manifests itself as "motor tremor," a cyclic variation in a constant, applied muscular force. Tremor has certain characteristic frequencies, but itself shows variation. The additional noise responsible for this variation might be related to probabilistic patterns in the firing of neurons in the central nervous system (on the source of which see Strevens [2003, section 4.94]), but there are difficulties in establishing such a connection (Todorov 2002, 1253–1254). This is a topic to revisit, perhaps, in a decade or two.

Whatever its ultimate biological basis, the across-the-board smoothing effect of neuromotor noise supplies a platform for microequiprobability in a wide range of human action, including the croupier's characteristic moves. On this physical fact an epistemological rule can be built: assume the microequiprobability of human movement, relative to some physiologically determined standard for what is microsized—which perhaps proportions what is "micro" to the magnitude of the move.

How does this rule relate to the wider-scoped microequiprobability rule? First, it provides more specific guidelines about the standard of "micro" than that rule.

Second, it provides a more secure foundation for the assumption of microequiprobability: for human movements, you can be more confident of microequiprobability than for changes in position generally; in particular you know that there are few if any likely defeaters. When you can, then, you should apply the narrower but surer rule, much as in the case of scenarios in which the microdynamic rule applies. Your conclusion will be the same as if you had applied the wider rule, but it will arrive upon a surer footing.

What is true of humans is presumably true of many other animals. There is perhaps, then, a *physiological microequiprobability rule* imputing microequiprobability to the distributions of animal movements generally.

Does the physiological microequiprobability rule come with custom defeaters? To undercut the application of the rule to, say, prestidigitators, card sharks, and sharpshooters? Perhaps, but it is unclear that customization is necessary. Nonmicroequiprobabilistic statistics are a defeater for the microequiprobability rule, and to learn of the existence of prestidigitators and so on is to learn of the existence of individuals capable of reliably producing such statistics. Broad epistemic considerations are sufficient, then, to prevent the dimwittedly dogged application of the physiological rule to these talented individuals.

You might imagine that there are many other domains besides human action where dynamic facts license the ascription of microequiprobability to some broad class of variables. Gather these special cases together, and they might constitute a form of the microequiprobability rule itself, founded not on the top-down considerations to be given in section 12.3 but on a bottom-up construction that sews together many patches of microequiprobability to produce something approaching a blanket assumption. For this mosaic rule to have psychological reality as well as epistemic force, we would have to be capable of recognizing the patches for what they are; whether we have such a compendious knowledge of the many tranches of microequiprobability, I do not know. At least in the case of humans' and perhaps other animals' movements, however, I surmise that we do recognize the special circumstances warranting increased confidence in microequiprobability.

12.3 Perturbed and Disturbed

Microequiprobability is everywhere—not just in the local burg or biosphere, not just among things that think and bleed, but everywhere. Why? Because perturbations, the disturbances that generically make up environmental noise, tend to smooth out initial conditions. Or so I will try to convince you.[7]

My explanation presupposes that the initial conditions in question— the conditions whose microequiprobabilistic distribution is to be predicted or explained—are themselves produced by a causal process with its own initial conditions, which I call the *ur-conditions*. Each initial velocity of a die roll, for example, is produced by a causal process rooted in croupier physiology, from a set of ur-conditions that presumably include physiological states, relevant environmental conditions, and so on. Assume that the initial conditions falling under the probability distribution whose microequiprobability is to be explained are created by a single type of generating process; this might be regarded as a simplifying assumption, or as a criterion for individuating initial condition distributions (Strevens 2011). I do not assume that the ur-conditions are uncaused: they may themselves have been produced by earlier causal processes from even ur-er conditions.

The perturbation explanation of microequiprobability turns on the idea that environmental noise or other external disturbances affecting the initial-condition generating process act so as to flatten to some degree the distribution of initial conditions, creating microequiprobability if not full equiprobability. (It may be that the initial conditions would have been microequiprobable anyway, in which case the perturbations may have no further net smoothing effect; however, it will be simpler to write in the active idiom.)

Distinguish two kinds of perturbation of the initial-condition generating process, noise and bias. (Other possibilities will be countenanced briefly below.) Noise consists in multitudinous small disturbances of the generating process, each throwing some particular instance of the process slightly off course (relative to its trajectory in a noise-free environment). The impact of air currents on a tossed coin or an errant swarm of insects on a foraging deer are examples. Bias consists in a more systematic, but still accidental or haphazard, influence on the generating process, affecting in the same way a large number of instances of the generating process. The impact of a croupier's temporary excitement on their rolling of a die or of a predator's accidental ingestion of fly agaric on its level of aggression are examples. While it is natural to think of a piece of noise as knocking the system off the trajectory prescribed for it by the evolution function, it is natural to think of a bias as a temporary perturbation of the evolution function itself (though there is in philosophy of course no proscription against unnatural thoughts).

Consider the effect of noise in the generation of initial conditions. A particular environmental fluctuation will have the effect of perturbing the production of a single instance of an initial condition. Call this instance the outcome condition. At some point, the fluctuation will give the production process a slight nudge, as a result of which the outcome condition will have a value somewhat different from the value it would otherwise have had. I want to investigate the effect of these perturbations on the distribution of outcome conditions. I assume that, in the absence of noise, there is a well-defined probability distribution over the outcome conditions, and ask how, if at all, that distribution is altered once noise is introduced.

Make the following assumptions about noise and its effects (noting the distinction between the noise itself and the ultimate effect of the noise on the outcome conditions, which is determined by the noise together with the dynamics of the process perturbed by the noise):

1. Noise is probabilistically distributed: there is some probability distribution over the magnitude and direction of individual fluctuations.
2. The probability distribution over noise is such that individual fluctuations are microequiprobable, independent of one another, and independent of the value of the ur-condition for the perturbed process.
3. The dynamics of the initial-condition generating process are such that the resulting perturbations of the outcome conditions have the same properties: they are microequiprobabilistically distributed, independent of each other, and independent of the value that the outcome condition would have taken if unperturbed (which I assume is determined by the ur-condition). In the jargon, the distribution of perturbations of the outcome conditions is IID, independent and identically distributed.

Under these assumptions, the noise will have a strong probabilistic tendency to smooth out the distribution of outcome conditions, so that a nonmicroequiprobabilistic distribution very likely becomes microequiprobabilistic, while a distribution that is already microequiprobabilistic very likely remains that way.[8] This is the simple version of the perturbation explanation.

An informal proof: the distribution over outcome conditions that results from the imposition of noise on the noise-free distribution is a weighted sum of many overlapping "copies" of the IID distribution over outcome perturbations—one copy for every possible ur-condition, centered on the outcome condition produced by that ur-condition when unperturbed and weighted by the ur-condition's probability. (Each such copy represents the probability distribution over outcome conditions resulting from a specific ur-condition plus noise.) A weighted sum of many microequiprobabilistic distributions is itself microequiprobabilistic, since its form over any microsized interval is the weighted sum of many approximately uniform distributions over that interval. The argument applies as well to joint distributions, whose microequiprobability is also assured by the microequiprobability rule (note 2).

How reasonable are the assumptions required by the perturbation explanation? Rather than giving a comprehensive answer, let me pick out what I see as the assumptions' two most insecure aspects.

First, it seems unlikely to be true in general that the size or direction of fluctuations will be completely independent of the value of the ur-condition affected. The independence assumption can be weakened, however: all that is necessary is that the joint distribution over fluctuations and ur-conditions affected is microequiprobabilistic. This allows that the size and direction of a fluctuation is correlated with the *approximate* value of the ur-condition. (In the terminology of Strevens [2003], the fluctuations may be correlated with high-level information about the ur-condition, but not with low-level information.) Why is microequiprobability of the joint density enough? It is sufficient to entail that the distribution over the outcome conditions is the weighted sum of many microequiprobabilistic distributions,[9] even if it no longer entails that these distributions are identical. And the proof above requires only the weaker condition.

Second, in some cases it may seem unlikely that the distribution over environmental noise is microequiprobabilistic, just because the standard deviation of the distribution is less than microsized; that is, because fluctuations are tiny even by the prevailing standard of "micro." What really matters for the proof, however, is that the ultimate effect of the fluctuations on the outcome conditions is not tiny in this sense. If the dynamics of the process producing the outcome conditions is sensitive to initial conditions then small fluctuations will have much larger effects, so the relevant standard of "micro" for judging whether assumption (2) holds is much smaller, making the assumption considerably easier to satisfy.

What if it is, nevertheless, not satisfied? What if, in particular, the effect of noise on outcome conditions is distributed smoothly—making a bell curve, perhaps—but with a width or standard deviation that is less than microsized? Though noise of this sort has a smoothing effect, it is not sufficiently powerful to overcome extreme nonmicroequiprobability in the ur-conditions. The smoothing will, however, make extreme non-microequiprobability less extreme: the distribution of the outcome conditions will be closer to microequiprobability than the distribution of the ur-conditions. There is the prospect, then, of iterating the perturbation explanation, as follows. The ur-conditions themselves are produced by some generating process subject to environmental noise (or so the iterated explanation supposes). Thus they have already been smoothed somewhat, as have their own ur-conditions before them. Given a mild assumption of independence, the aggregate effect of these iterated smoothings will be greater than the effect of any single smoothing (because the variance of the aggregate noise is the sum of the variances of the individual noise distributions). With sufficient iteration, then, the aggregate smoothing will imply the microequiprobability of the outcome conditions, as desired.[10]

If these defenses are successful, the perturbation explanation's fundamental premise stands: environmental noise will tend to smooth out distributions of initial conditions, creating a trend to microequiprobability wherever there is something less than total silence. But does the explanation perhaps imply too much? As I showed you in section 12.1, for every random variable that is microequiprobabilistically distributed there is another, describing the same physical quantity, that is not microequiprobabilistic. How can all variables tend to microequiprobability, then?

They cannot. But this fact is accommodated by the explanation as it stands. Suppose that v and v^* are two such variables, as in figure 12.1. Suppose further that assumption (3) above holds for v: the dynamics of the process producing outcome conditions converts a microequiprobabilistic distribution over noise into a microequiprobabilistic distribution over perturbations of v. Then it is not hard to see that assumption (3) will fail to hold for v^*, so that the same distribution over noise results in perturbations of v^* that are not microequiprobabilistic.

It is this observation that is the key to understanding the special status of what I have called the "standard variables," the variables used by science to quantify physical magnitudes. The standard variables have the following property: physical processes tend to change their values in a certain sense rather smoothly, so that microequiprobabilistic distributions over inputs are translated into microequiprobabilistic distributions over outputs. Thus, assumption (3) is likely to be true for standard variables, but false for those rogue nonstandard variables whose microequiprobability is incompatible with some standard variable's microequiprobability.

What the perturbation explanation shows, then, is that standard variables have a tendency to get themselves distributed microequiprobabilistically; at the same time it explains why the most pathological nonstandard variables have precisely the reverse tendency. An objective rationale for the microequiprobability rule is thus provided.

Some clarifications. First, the smoothness in dynamics that distinguishes standard variables from the most pathological nonstandard variables is what I called in section 7.1 *microlinearity;* you will recall that a process is microlinear relative to a variable if its effect on the variable is approximately linear over any microsized interval.

Second, if a standard variable is microequiprobabilistically distributed, then many nonstandard variables quantifying the same physical quantity are also microequiprobable, notably, those nonstandard variables that are microlinear transformations of the standard variable (holding the physical significance of "micro" fixed). Standardness is a sufficient condition for a variable's having a tendency to microequiprobability, then, but nowhere near a necessary condition. (Note also that in those cases where a quantity can be represented in more than one standard way, such as the Cartesian and polar representations of velocity, the corresponding sets of variables are each microlinear functions of the other.)

Third, there is no physical prohibition against a standard variable undergoing a nonmicrolinear transformation. This is why it is possible, even easy, to build a device that transforms a microequiprobabilistic distribution over a standard variable into a nonmicroequiprobabilistic distribution over the same variable.[11] The assumption of microlinearity that stands behind assumption (3)'s holding for standard variables, then, should be understood as a contingent claim about typical physical processes "in the wild": most such processes are microlinear.

The same goes for the assumption, a couple of pages back, of sensitivity to initial conditions. Some processes might be so insensitive that there is almost no variation in their output—if, for example, a quantity is maintained by extremely precise and reliable homeostasis, or if it has such great inertia that noise has no discernible impact on its values. The reliability of the microequiprobability rule depends on such processes being either rare or easily recognized.

Fourth, the perturbation explanation invokes contingent facts about the distribution of noise (at least, if the world is at root deterministic), namely, the facts underlying assumptions (1) and (2), or better, the weaker version of the assumptions that does not assume an IID distribution of perturbations. The whole thing depends, then, on our world being a certain way that it did not have to be. Equidynamics need not worry about unrealized spoiling possibilities, however; its reliability hinges on what is, not on what merely might have been.

Fifth, in the course of developing the perturbation argument I have drawn on both sensitivity to initial conditions and microlinearity, and used them to derive microequiprobability in roughly the same way as in the vindication of the microdynamic rule in section 7.1. Although the rules' justifications are mathematically very similar, they are quite different in form: one requires the user to verify the presence of the relevant dynamic properties, but the other does not. This is why the conclusions warranted by the microdynamic rule are more secure.

The microequiprobability rule allows you to assume the approximate uniformity of a probability distribution over a microsized interval. But what is "microsized"? The perturbation explanation supplies an objective physical basis for an answer: the standard for "micro" is determined by the magnitude of the perturbations' smoothing effect. The standard will vary as the smoothing dynamics vary, from variable to variable and from context to context.

Do our equidynamic smarts extend to making fine judgments of this sort? I am not sure. Perhaps we are responsive, in cases that are not wholly epistemically exogenous, to the generating process's degree of microlinearity and inflation, which put an upper bound on the standard of "micro" relative to which perturbations create outcome condition microequiprobability. If not, the perturbation argument is of little help to us in determining the standard; we will have to fall back on situational knowledge or even a general rule of thumb such as the "2 percent rule" mooted in section 12.1.

There is a hole in the perturbation explanation. It shows that what I have called noise will tend to create microequiprobability, but it does not rule out the possibility that other equally powerful environmental influences will tend to undermine microequiprobability. Consider, for example, what I called at the beginning of this section "bias": a systematic but short-term influence on the production of outcome conditions. Can bias counteract the smoothing effect of noise?

The plausibility of the perturbation explanation depends in part on the supposition that the dynamics producing initial conditions is microlinear. If a bias were to undermine this microlinearity, the smoothing effect of noise might be disrupted. But equally, if the bias does not undermine microlinearity, it will have no effect on the tendency to microequiprobability. If most biases can be thought of as tweaking the parameters of a dynamics that is essentially microlinear—for example, increasing the mean speed of a croupier's spin—then they for the most part do not endanger microlinearity, and so the perturbation explanation stands.

As for other possible underminers of noise's smoothing effect: as Hume said of justifications for induction, if you find one, let me know.

12.4 Noise Ignored

In past chapters—throughout part 2, in particular—my investigation of the connection between dynamics and probability has been lubricated by a simplifying assumption, that the outcomes of the processes to which equidynamics assigns physical probabilities are entirely determined by their intrinsic initial conditions, for example, that the outcome of a die roll is entirely determined by the die's initial position and velocity, along with the motion of the roller's hands. In reality, environmental disturbances, such as air currents and eddies or small movements of the table on which the die lands will affect its motion, if only slightly. By what right do I ignore this noise?

A piece of noise—an environmental fluctuation—will sometimes nudge things one way, sometimes the other. Will these fluctuations bias the probability distribution over the process's outcomes? To do so, they would have to, on average, push the system toward a given outcome more often than they deflect it from that same outcome. Put in terms of an evolution function, they would have to push a system's trajectory into one of the "gray areas" (the areas corresponding to a given outcome's occurrence, literally colored gray in figure 5.1 and elsewhere) more often than they deflect the trajectory from the same areas.

Such a biasing effect can exist only if there is a correlation between the direction and magnitude of the "nudge" and the precise details of the initial conditions, what Strevens (2003) calls the "low-level information" in the conditions. If there is such a correlation, then the joint probability distribution over the fluctuations is not microequiprobabilistic. Thus if the fluctuations are distributed microequiprobabilistically, noise has no biasing effect. (Similarities to the explanation of microequiprobability in section 12.3 are not accidental.)

Noise can be ignored, then, if the resulting fluctuations tend to be microequiprobabilistic. When we do ignore noise, we are therefore making what I called above an implicit use of the microequiprobability rule.

12.5 Probabilities and Tendencies

Equidynamic reasoning, I have so far supposed, assigns physical probabilities only to those outcomes produced by causal processes whose initial conditions have physical probability distributions. As its name suggests, equidynamics for the most part focuses on epistemically leveraging the properties of the processes. The physical probability distribution nevertheless serves two auxiliary roles. First, by being microequiprobabilistic, it licenses the application of methods such as the stirring rule to infer probabilities from dynamics. Second, by being, *simpliciter*—by existing at all—it gives these methods a subject matter, by founding the existence of physical probabilities whose values equidynamics may then determine.

The second of these roles raises parallel metaphysical and epistemological questions:

1. Must there be a physical probability distribution over initial conditions for there to be a physical probability distribution over outcomes?
2. Must there be a physical probability distribution over initial conditions for equidynamics to yield reliable information about patterns of outcomes?

Let me begin with the latter, epistemic question.

Consider some novel gambling device, such as a complex but regularly proportioned three-dimensional object treated as a die—maybe the dodecahedron from chapter 3 (figure 3.1), with one of its twelve sides painted red. The dynamics of this regular, twelve-sided object compel an equidynamic judgment that the probability of the outcome *red*—the event of the red side's ending uppermost after a toss—is one in twelve. What needs to be true of the initial conditions if this probability is to have predictive value, that is, if the frequency of *red* is to be approximately 1/12? Or—let us keep an open mind as to what constitutes a predictive probability—to be "very likely" 1/12, or 1/12 with high probability?

A condition for predictivity suggests itself at once: the actual initial conditions of the die rolls should be smoothly distributed, in roughly the same sense of "smooth" characterized by the definition of microequiprobability. What does it mean for a distribution of actual initial conditions—presumably finite, or at most denumerable, in number—to be "smooth"? Divide the range of the initial condition variables into equal-sized, very small (less than microsized) intervals, and plot a histogram recording the number of conditions falling into each interval. If neighboring bars in the histogram are for the most part of roughly equal height, you have a smooth distribution of actual frequencies. (For a slightly more sophisticated notion of smoothness, using a "frequency density" that represents a moving average, see Strevens [2003, section 2.33].)

Some brief remarks. First, such a definition of smoothness will be useful only if there are many actual initial conditions to give shape to the histogram. Second, for predictivity, smoothness of the distribution of actual initial conditions is a sufficient condition only; any number of jagged distributions of the actual initial conditions would yield an approximate 1/12 frequency for *red* "by chance," as it were. Third, for predicting events more complex than the frequency of *red*, such as the frequency of *red* on both of two consecutive rolls, the corresponding condition would require the smoothness of the joint distribution of sets of actual initial conditions (see section 5.3). For expository simplicity, I will not talk explicitly about joint distributions in what follows (except in note 13), leaving you to fill the lacunae at your leisure.

To repeat, smoothness of the actual initial conditions is sufficient for the predictivity of the 1/12 probability of *red*, and by extension for the predictive success of equidynamics more generally. You might, then, looking to entrain the definition of physical probability to the predictive success of equidynamics, stipulate that there exists a physical probability for *red* provided that

1. the dynamics of the die roll, and in particular the evolution function for *red*, are microconstant, and
2. the actual initial conditions of die rolls are distributed smoothly,

with the value of the probability set equal, of course, to the strike ratio for *red*.

Such a definition gives equidynamics a physical probability to infer, and thereby a suitable metaphysical stepping stone to the prediction of the correct frequency, which assumes only the existence of frequentistic, and not probabilistic, facts about the distribution of initial conditions.[13]

You might think that this stepping stone is insufficiently ontologically exalted to qualify as a genuine probability; it is a mere "as if" probability, or as you might call it, an "alsobability."[12] That is fine; you can then think of equidynamics as a technique for inferring the values of those marvelous predictive vehicles, the alsobabilities. It is an open question, which I have no plan to resolve here, whether the facts underlying Maxwell's, Darwin's and others' scientific contributions are genuinely probabilistic, or only alsobabilistic, facts.

So then: equidynamics can be invoked profitably for predictive purposes in some cases where there is no physical probability distribution over initial conditions, regardless of whether you think that the prediction goes by way of a real physical probability or an ersatz alsobability. Can the same be said for explanation? Can the probabilities (or alsobabilities) not only predict, but also explain, the 1/12 frequency of *red?*

That depends on your theory of explanation. Elsewhere (Strevens 2008, chap. 10) I have offered an account of explanation on which the smoothness of the die's actual initial conditions on some long series of rolls, together with the relevant properties of its dynamics (its microconstancy with a strike ratio for *red* of 1/12), do explain the resulting 1/12 frequency of *red.* So far, so good.

The same frequency may also be explained by a slightly different set of facts, according to my account. In this alternative explanation, the dynamic facts—the facts determining microconstancy and the strike ratio of 1/12—remain the same, but instead of the initial conditions' actual smoothness, the explanation cites their *tendency* to smoothness, if this tendency exists.

A tendency to smoothness in the explanatorily relevant sense exists when a sufficiently large set of counterfactual conditionals of the following form holds:

Had the series of rolls been conducted under slightly different conditions, the initial conditions would likely (still) have been smoothly distributed.[13]

Each of the conditionals in such a set counterfactually supposes a distinct "slight difference"; some examples would be "Had the rolls been conducted one minute later . . . ," "Had the die been slightly heavier . . . ," "Had the rolls been more widely spaced . . . ," and so on.

There are two ways to interpret the "likely" that appears in the counterfactuals (assuming a possible worlds semantics). On the first interpretation, "If *A* then likely *B*" means that, in the closest possible worlds where *A* occurs, *B* has a high physical probability. On the second interpretation, "If *A* then likely *B*" means that, in *most* closest possible worlds where *A* occurs, *B* also occurs, where the *most* presupposes the existence of a measure over possible worlds that is built in to the semantics of "likely" counterfactuals.

According to the first interpretation, the initial conditions of a process have a tendency to smoothness only if they are induced by a counterfactually robust physical probability distribution.[14] According to the second interpretation, by contrast, a tendency to smoothness can exist in a set of initial conditions that has no prospect of falling under a physical probability distribution. If the second interpretation is allowed—if, as Strevens (2008) argues, the truth of the relevant "likely" conditionals interpreted in the second way together with a microconstant dynamics is sufficient for the explanation of frequencies—it follows that equidynamics can tell us about probabilities (or alsobabilities) that are both predictive and explanatory even in the absence of a physical probability distribution over initial conditions.

The details of the second kind of "likely" conditional and its basis in nonprobabilistic facts (or at least, in facts that imply no physical probability distribution over initial conditions) are discussed in Strevens (2011). The same paper formulates and argues for a definition of physical probability based on initial conditions' tendency to distribute themselves smoothly.

The microequiprobability rule requires for its application the existence of a physical probability distribution over initial conditions. Is there some expanded or altered version of the rule that tells us when it is reasonable to infer the existence of a tendency for the actual initial conditions to become smoothly distributed even where there is no physical probability distribution driving the tendency?

You might assemble a mosaic rule from more specific rules that cover particular classes of initial conditions, as described in section 12.2. One piece of the mosaic might tell you that the frequencies of human muscular motions tend to be smoothly distributed, another that the actual sizes of animals tend to be smoothly distributed, and so on.

Or you might try for a fully general rule. You could build the rule around the perturbation explanation: assume a tendency to smooth statistics when some or all of the dynamic conditions assumed by the perturbation explanation are satisfied.[15] The rule might in that case have roughly the following form:

> Defeasibly assume that the actual values of any variable show a tendency to become smoothly distributed provided that (a) it is a standard variable, (b) its instances are produced by a causal process, (c) the initial conditions and/or the process are subject to perturbations or environmental noise, (d) the perturbations are of a scale sufficient to flatten out distributions over microsized intervals.

The mass of protons fails to satisfy the rule's conditions of application because either (b) or (c) is violated (depending on how the physics of proton mass turns out). The sizes of *Ibla cumingi* barnacles fail to satisfy the rule's conditions of application because (d) is violated—there being no matter of fact about the size of the appropriate scale, since the entire range of variation in the sizes of the males would count as only a microsized interval relative to the variation in the sizes of the females. You might require as a further condition on the application of the rule that the initial conditions in question are all produced by the same kind of causal mechanism. That would provide a further reason to place *I. cumingi* size outside the rule's scope. No doubt further refinements are necessary, but I am stopping here.

13

THE ELEMENTS OF EQUIDYNAMICS

The various inference rules that drive equidynamic reasoning in adult humans, postulated, exemplified, and applied in previous chapters, will here at last be assembled as an organic whole.

13.1 The Structure of Equidynamics

The rules of equidynamics may be divided into those that have no dynamical premises (that is, no premises about physical processes) and yield conclusions about physical probability distributions over initial conditions, and those that have dynamical premises and yield conclusions about physical probability distributions over outcomes. The root of the distinction lies in the nature of the premises, since the difference between an initial condition and an outcome is, extreme points aside, perspectival. Call the former rules, nevertheless, the *initial condition rules;* call the latter rules the *dynamic rules.*

Of the dynamic rules, the simplest in function are the rules of proportional dynamics, which calculate the proportion of some set of initial conditions leading to a given outcome, in either absolute or relative terms. The calculation need not hinge on physical generalizations in the narrow sense. Biological or psychological generalizations will underlie proportional dynamic thinking about organisms and minds. To put it another way, proportional dynamics may be guided by folk biology and folk psychology as well as by folk physics, or in yet other words, by biological and psychological intuition as well as physical intuition.

In many of the principal applications of equidynamics discussed above, proportional dynamics has played only an auxiliary role, helping us to determine, for example, the biasing effect of the magnetic claw (section

8.2) or the nonbiasing effect of a tossed coin's bounce (section 6.1). But it can, when used in conjunction with the initial condition rules, work a magic all its own, as chapters 9 and 11 show. I have not attempted to understand the foundations of proportional dynamic reasoning, a vast and important topic that will surely, as remarked at the end of section 10.2, cast much light on the scope and power of equidynamic reasoning.

The same can be said for the second kind of dynamic rule, the methods of probabilistic dynamics, which allow the user to calculate an entity's likely trajectories through the space of possibilities on the grounds of its stochastic microdynamics, as when we answer the question whether an urn has been shaken long enough to thoroughly randomize the position of a given ball. Probabilistic dynamics has independent uses, but it is especially valuable in the application of—though it is not a part of—the equilibrium rule package, where it is used to estimate relaxation time and to apply the uniformity rule.

A third dynamic rule applies to a special class of microconstant processes, the stirring processes, of which my principal examples have been the wheel of fortune and the coin toss. A fourth rule, the majority rule, leans heavily on proportional dynamics; it was applied to evaluate the selective advantage of the "freeze" strategy in section 9.3.

The equilibrium rule package is the fifth dynamic rule investigated in this book. It drives our equidynamic reasoning about what I have called the shaking processes, processes in which a "shaking parameter" randomly visits its space of possible states, with its value at some predetermined time—the moment that a die is released from the hands or a ball drawn from the urn—fixing the outcome of the process as a whole. Besides the die and urn, many other familiar gambling setups depend on shaking for randomization, most prominently perhaps the roulette wheel.

In much the same way, the equilibrium rule package guides our reasoning about a range of processes that do not in any natural sense have a single focal outcome, such as balls careening around a box, animals cavorting around a habitat, and so on up the great hierarchy of being. (This expositorily useful distinction between processes with and without a focal outcome is, I should add, a shallow one: focus can be added to a box of bouncing balls by cutting an aperture and waiting for a ball to exit, or simply by designating a focal ball, waiting a certain amount of time, and noting whether that ball is in the left or the right side of the box.)

A few reminders. First, I conjecture that the rules of equidynamics are universal—that is, that they are used by all mentally competent adult humans (and perhaps even very young children). My evidence for this is limited. Perhaps the most striking point in favor of universality is the appearance of equidynamic reasoning in young children, which suggests that central facets of equidynamics are innate. Universality matters more to me, however, than innateness.

Second, equidynamic rules are to be applied with respect to standard variables only; standardness matters because the standard variables, and other variables that are microlinear functions of the standard variables, tend to distribute themselves microequiprobabilistically. In this way equidynamics avoids the ambiguities made famous by Bertrand, since the equidynamic rules applied with respect to the standard variables cannot give conflicting advice. A stronger claim is also true: As explained in Strevens (2003, section 2.5), any variable with respect to which the equidynamic rules make prescriptions that conflict with their prescriptions relative to the standard rules, is a variable that has a tendency to distribute itself nonmicroequiprobabilistically (see figure 12.1). So the equidynamic rules applied relative to the standard variables not only give good advice, but uniquely good advice, on the assumption that goodness of advice requires a tendency to microequiprobability in the variables in question.

Third, what you might naturally count as a single process may be broken down by equidynamic reasoning into several subprocesses to each of which a different rule is applied, as when a die roll is broken down into setting (positioning the die in the hand), shaking, and the roll itself.

Fourth, equidynamic reasoning is defeasible. The simplest defeater is statistical information: if the frequencies do not reflect the inferred physical probabilities, the equidynamic inference that supplied those probabilities may be canceled. If the rule in question is dynamic, some physical premise of the inference is most often blamed, as when Karl Pearson concluded from the slightly irregular statistics of 26,306 rolls of twelve dice, conducted in 1894 by the biometrician Raphael Weldon with the help of a university clerk, that Weldon's aleatory tools were imperfectly cubical (Pearson 1900).[1]

Fifth, the conditions required for many of the equidynamic reasoning rules to deliver reliable verdicts are stronger than the conditions under which we are disposed to invoke them. In principle, then, some of our applications of the rules will be unreasonable, in the sense that information about the case in question is available suggesting—to a sophisticated scientific mind, at least—that the rule is unlikely to deliver accurate judgments. We will, in short, overgeneralize about physical probability. I know of no examples of overgeneralization, however, which suggests that the circumstances in which equidynamic reasoning goes wrong are rare.

Sixth, I have no reason to think that the suite of equidynamic rules described in this chapter is complete. Indeed, they cannot be completely complete, because of the imprecision in some aspects of their formulation—in what degree of sensitivity to initial conditions is required to trigger the microdynamic rule, for example.

The details can be filled out by experimental psychology, perhaps, but not by the kind of arguments presented in this book, which have relied on three kinds of evidence to establish that the proposed rules guide our thinking about physical probabilities:

1. The rules explain a good proportion of our equidynamic inferences, including sophistications such as the role of relaxation time in reasoning about urn drawings. They explain also why certain probabilistic models, without statistical confirmation of any sort, have had sufficient plausibility to alter the course of science.
2. The rules have a high degree of ecological validity, meaning that they are reliable given the typical workings of the world in which they are applied.
3. The rules have a certain cognitive resonance: they appeal to the kinds of considerations that seem, intuitively, to be relevant to determining physical probabilities. They are, in other words, familiar in the way that you would expect from something that has been living quietly all these years in your brain.

Such considerations put powerful constraints on hypotheses about equidynamic reasoning, but their power is in large part due to a broad qualitative sweep that cannot be expected to resolve the details with extraordinary precision.

For the same reason, the proposed rules are no more likely to be exactly correct than entirely complete. Indeed, some aspects of the rules formulated here are frankly speculative; regard them as a first step on the path to the equidynamic truth.

The next few sections exist to summarize for your convenience what I have said in earlier chapters about the initial condition rules, the stirring rule, the majority rule, and the three parts of the equilibrium rule package.

13.2 The Initial Condition Rules

Assume that the physical probability distribution over any quantity, as represented by standard variables, is microequiprobabilistic. Assume, more exactly, that the joint distribution over any collection of instantiations of the quantity is microequiprobabilistic (see note 2 of chapter 12), meaning that it is approximately uniform over any contiguous microsized region, with the independence properties that this implies. So says the microequiprobability rule, the most sweeping of the initial condition rules.

Stated thus, there are no preconditions whatsoever for the rule's application, apart from the existence of a probability distribution to which to apply it, though like all equidynamic rules it is subject to defeat by statistical and other information (section 12.1). Even a probability distribution may not be required, I suggested in section 12.5: a more liberal version of the microequiprobability rule might allow you to infer a tendency to microequiprobability in a quantity's actual instantiations, which would have predictive and explanatory power with or without the oversight of a governing physical probability distribution.

What counts as a microsized interval? I have not given a definitive answer to this question. Perhaps there is some universal, default standard for "micro" built into our system of equidynamic reasoning, such as section 12.1's "2 percent rule." Or perhaps dynamic considerations lend a hand: what is micro is to be determined by the average size of the perturbations whose smoothing action creates microequiprobability and so provides the rule's foundation. If so, the rule is not entirely free of dynamic elements.

There are versions of the microequiprobability rule that apply to special cases. Most salient in this book has been the physiological microequiprobability rule, which licenses the assumption that quantities generated by human action (and perhaps the actions of other animals) have a microequiprobable probability distribution. What does this add to the more general rule? First, when the specific rule applies, the conclusion may be inferred with greater confidence than in the general case. Second, the specific rules presumably offer more precise guidance as to what counts as "micro."

Another initial condition rule, discussed only in passing, is the familiar rule, which I call the *independence rule,* licensing the assumption that two outcomes are stochastically independent if they are produced by causally independent processes. The independence rule is more subtle than it appears, I think: on a strict interpretation of causal independence the principle is correct but, in a universe such as ours where the causal history of everything goes back to a single event of cosmic initiation, it has no instances, while on an interpretation of causal independence weak enough to make the principle useful, it cannot be applied indiscriminately and so requires a further rider (Strevens forthcoming). Where the causally independent processes are microconstant, stochastic independence can more confidently be assumed—it can be assumed, in particular, that causally independent stirring and shaking processes produce stochastically independent outcomes—though this rule, turning as it does on the dynamic property of microconstancy, does not strictly speaking belong under the heading of initial condition rules.

All three of the initial condition rules mentioned here may involve qualitative dynamic considerations, I should note; even the physiological and independence rules require some knowledge of the processes creating the initial conditions in question. Perhaps, then, all equidynamic rules are dynamic to some degree. Still, the rules described in this section require for their use far less of a grip on the dynamics than the rules to be described in what follows.

13.3 The Stirring Rule

A stirring process is characterized by an outcome-determining quantity called the stirring parameter; stirring happens when the process cycles smoothly through every value of the stirring parameter sufficiently many times (in the senses of "smoothly" and "sufficiently many" characterized in section 5.4). The stirring rule allows us to infer that the outcomes produced by a stirring process fall under a physical probability distribution with the following properties: (a) the probability of an outcome is equal to the proportion of values of the stirring parameter yielding that outcome, (b) the outcomes are stochastically independent.

The rule is reliable only because the initial conditions of stirring processes tend to be microequiprobable. It does not, however, require microequiprobability of the initial conditions as a premise, though nonmicro-equiprobability may function as a defeater.

13.4 The Majority Rule

Suppose that one of two events must occur. Suppose that a considerably greater proportion of initial conditions lead to one event's actualization than to the other's. Suppose also that which of the two events occurs is sensitive to initial conditions: for every set of initial conditions leading to the occurrence of one of the events, there is another set nearby leading to the occurrence of the other. Then given the microequiprobability of the initial conditions, you may infer that the one event is more probable than the other. That is the absolute version of the majority rule.

A relative version, which tests a causal factor for probabilistic relevance to an outcome, was invoked in section 9.3 to explain our reasoning that it is under a broad range of circumstances advantageous to freeze in the proximity of a predator. The rule licenses you to infer that the factor in question increases the probability of the outcome if, in the factor's presence, many more initial conditions lead to the outcome than in its absence. As with the absolute version, this rule turns on sensitivity to and the microequiprobability of the relevant initial conditions.

A converse version of the relative rule may also find a place in our equidynamic inventory, licensing you to infer that a causal factor is probabilistically irrelevant to an outcome when the addition of the factor makes no difference to the proportion of initial conditions leading to and away from the outcome. There may be more sophisticated rules than this for inferring probabilistic irrelevance in such cases; if so, they will have to wait for volume two.

The majority rule is no sure thing: all premises of the inference may be true but the conclusion false (note 13 of chapter 9). The conditions under which the rule fails are, however, rare.

Does the rule require its user to make explicit assumptions of initial-condition sensitivity and microequiprobability? A cautious yes to both. But I cannot rule out the possibility that the microequiprobability assumption is implicit, in the same way that it is in the stirring rule, in which case the application of the majority rule does not require its user to have positive reasons for supposing initial-condition microequiprobability—or even that it should cross their mind.

13.5 The Microdynamic Rule

If an entity, such as a bouncing ball, has its movement through its space of possible states dictated by a series of interactions with other objects, such as other balls or container walls, and these interactions satisfy the sensitivity and smoothness conditions laid out below, then infer a micro-equiprobabilistic distribution over any aspect of the entity's post-interaction state, that is, over any of the entity's state variables. As well as microequiprobability in the narrow sense—uniformity of the state variable's probability distribution over any microsized region—infer that the value of the variable within a given microsized region (e.g., in the case of position, microposition) is independent of all other state variables in the system, including values of the same state variable at times that precede its most recent interaction. For example, when position is one of the state variables, infer that an entity's microposition after a given interaction is uniformly distributed independent of the entity's other properties, such as its direction of travel, independent of other entities' properties, and independent of its own pre-interaction state, including its pre-interaction position.

The sensitivity condition requires that the entity's post-interaction state depend sensitively (of course) on its pre-interaction state; the smoothness condition requires that the interaction dynamics be microlinear, that is, approximately linear over small regions of initial condition space.

It is important here as elsewhere, I remind you, to parameterize the interactions using standard variables; the microdynamic rule requires that the sensitivity and smoothness conditions be satisfied with respect to at least one standard quantification of the entity's state (in which case, I surmise, the conditions will be satisfied with respect to any standard quantification).

If some but not all aspects of an entity's post-interaction state depend sensitively on its pre-interaction state, the microdynamic rule may be applied by way of the following procedure (section 9.4):

1. Divide the entity's state variables, reparameterizing (standardly) if necessary, into two sets, so that the post-interaction values of the variables in one set depend sensitively on their pre-interaction values for every possible (or realistic) assignment of values to the variables in the other set.

2. Infer a microequiprobabilistic joint distribution over the variables in the former set.

If it is impossible to carry out the first step of the procedure, the microdynamic rule cannot be applied.

13.6 The Equilibrium Rule

The equilibrium rule warrants the conclusion that some property (represented by a state variable) of a given kind of entity in a given system may, over time, be regarded as executing a random walk that takes it to a unique long-term probability distribution—a globally stable equilibrium distribution. The rule may also warrant the conclusion that the states of distinct entities are, in the long term, stochastically independent. Three conditions must hold for a state variable to fall under the rule's jurisdiction: randomization, boundedness, and relaxation.

Randomization requires the existence of a "stochastic variable" on which changes in the state variable depend sensitively. (Some aspect of the state variable itself may fulfill this function; for a bouncing ball, microposition is a stochastic variable.) The variable's stochasticity derives from its falling under a fixed physical probability distribution with a sufficient degree of independence. The equilibrium rule does nothing to establish the existence of this short-term distribution. When the rule has been applied in previous chapters, it is the microdynamic rule that supplies knowledge of the distribution, but in principle we might infer the existence of the distribution in other ways, for example, from quantum physics.

The boundedness condition requires that for all practical purposes, the state variable have certain minimum and maximum values.

Relaxation requires that the system evolve for a sufficiently long period; another way of recognizing this condition is to say that the equilibrium rule guarantees that a system will reach its equilibrium distribution only in the long run. Accompanying the relaxation condition is an intuitive picture of the equilibrium rule's rationale: as time goes on, the random walk induced by the stochastic variable dissolves the influence of the system's initial conditions, so that the distribution of the system's parts comes to depend only on the parameters of the walk.

The practical use of the equilibrium rule depends on there being some way to estimate the duration of the relaxation period. For this purpose I have attributed to the rule's users—ourselves—a certain degree of expertise in probabilistic dynamics, which allows us to determine how long a random walk must go on for the initial state of the walker to be "washed out."

13.7 The Uniformity Rule

Some state variable falls under the equilibrium rule; you should therefore infer that it has a global, stable equilibrium probability distribution. You should go on to infer that the equilibrium distribution is uniform (with independence of the states of distinct entities) if the variable passes a further test imposed by what I call the uniformity rule, and equally, you should infer that the distribution is nonuniform if it fails the test.

The test in question putatively imposes a uniform probability distribution over each of the system's state variables (using practically determined upper and lower bounds where necessary). The tester then uses probabilistic dynamics to forecast the short-term behavior of these distributions as the system evolves. If a variable's distribution remains uniform (and it falls under the equilibrium rule), a corresponding uniformity in its equilibrium distribution should be inferred, and vice versa.

Suppose that a state variable falls under the equilibrium rule but fails the uniformity test, meaning that its distribution deviates from uniformity under the conditions set by the test. By way of what I call the corollary to the uniformity rule, you may infer that the variable's equilibrium distribution will differ from the uniform distribution in a way that broadly reflects the nature of the short-term divergence displayed in the uniformity test.

13.8 Equidynamics and Symmetry

What happened to symmetry? In the principle of indifference, it was paramount. When I divided the principle into its epistemic and physical parts, the allegiance to symmetry was apparently inherited by both: one prescribed symmetries in epistemic probability reflecting evidential symmetries; the other prescribed symmetries in physical probability reflecting physical symmetries. The elements of equidynamics laid out above make no explicit appeal to symmetry, however. How could that be?

Physical symmetries have an important role to play in equidynamic inference, but they enter into such reasoning indirectly rather than being specified directly by the rules of equidynamics. It is in virtue of symmetries, for example, that a state variable falls under the uniformity rule. The smoothness required by the microdynamic rule is secured by an absence of extreme asymmetry. The regular visitation rates definitive of a stirring process are induced by physical symmetries. There are obvious connections between proportional and probabilistic dynamics and Laplace's talk of outcomes being "equally possible" or Leibniz's talk of outcomes occurring "equally easily."[2] And so on.

Symmetry is, all the same, less important to equidynamic reasoning than an exclusive focus on simple gambling setups such as tossed coins and rolled dice would suggest. As observed in section 7.4, the perfect roundness of bouncing balls, and in particular Maxwell's equiprobability of rebound angles, turns out to be irrelevant to the balls' most significant probabilistic properties; it is not necessary for their conforming to Maxwell's velocity distribution, for example. And equidynamic reasoning in evolutionary biology makes almost no use of organismic symmetries (although environmental isotropies do matter). Equidynamics is less about inferring probabilities from symmetries, then, than it is about inferring probabilities from dynamic properties in general. If one of these properties were to be elevated to the supreme station, it would be not symmetry but initial condition sensitivity.

13.9 Our Probabilistic Expertise

Humans are famously hopeless at reasoning about probability (Kahneman et al. 1982). We ignore base rates systematically, if not exceptionlessly; for example, in many circumstances we take no account of the prevalence of a disease when determining how likely a person is to have the disease given a positive test. We fumble simple principles of probability mathematics, taking a conjunction to be more probable than one of its conjuncts; in the canonical example, we think that it is more likely that Linda is a feminist bank teller than that she is a bank teller.

Humans are extraordinarily subtle and sophisticated reasoners about probability; that is the message of this book. From our earliest years, we are able to reason accurately about the stochastic dynamics of processes that are extremely difficult to analyze physically, such as urn samplings, or even nigh impossible to analyze physically, such as the goings-on in ecosystems. In such thinking, we seldom or never make avoidable mistakes: in particular, we are typically fully aware of the empirical suppositions on which our conclusions rest, and are ready to abandon the conclusions if those suppositions become doubtful, as we do if we learn that the die is loaded, that the claw sampling from the urn is magnetic (section 8.2), or that bees who feed on flowers with more nectar are easier targets for predators (section 9.5).

How can we be at the same time so brilliant and so bad with probability? The mix of expertise and ineptitude is not without precedent. Humans are naturals at applying *modus ponens*, but quite erratic when it comes to *modus tollens* (Holland et al. 1986). (You might detect an even closer parallel here, if you think of equidynamics as a kind of direct inference and the use of base rates as a part of inverse inference.)

Perhaps the most important thing to note, however, is that the probability treated by equidynamics is a rather different thing from the probability that we demonstrably mishandle: when thinking equidynamically, we are dealing directly with physical probability, whereas we are manipulating epistemic probabilities when reasoning about the significance of medical tests, guessing the occupations of mysterious strangers, and so on. Kahneman and Tversky's famous studies show that there are significant lacunae in our inductive reasoning abilities, but that is quite consistent with a domain-general genius at inferring physical probabilities.

14

PREHISTORY AND META-HISTORY

14.1 Equidynamic Darwinism

How do the elements of equidynamics get into our heads? The appearance of mature physical understanding in infants who have yet to take their first steps suggests, as I noted in chapter 4, a role for innate knowledge. Might our brains have been prepared to think equidynamically ahead of time? If so, the perpetrator would presumably be evolution by natural selection.

I am hardly in a position to offer an evolutionary explanation of the implementation of equidynamics in the human mind. But I can prepare the way by speculating intelligently about the advantages offered by equidynamic reasoning in the pre-agricultural wilderness. Some intelligence indeed seems to be needed: it is not immediately obvious how insight into the statistical behavior of tumbling polyhedra and bouncing spheres might have staked a place in the gray matter when there is so much demand for thought about biologically more imperative matters such as eating and mating. Our ancestors did not escape the saber-tooths by thrashing them at roulette. So what could be the point of burdening the primeval intellect with the microdynamic rule, the equilibrium rule, the microequiprobability rule, and the other inferential extravagances of equidynamics?

Let me give you several scenarios in which an equidynamic sophisticate might be at an advantage. Each is merely a just-so story, in the apologetic but not entirely derogatory sense of that term, and the whole is nothing more than brainstorming about possible pieces of an evolutionary explanation. Still, there is no better place to begin if you want to take seriously at all the possibility that equidynamic thinking has evolutionary origins.

Consider that saber-toothed tiger. Suppose that you are taking a shot at it with your spear. Should you aim for the head or the body? One important consideration: you are more likely to hit the body, because it is larger. Your reasoning here is identical to Keegan's reasoning about patterns of wounding in the Battle of the Somme, the equidynamic aspects of which, involving proportional dynamics and the physiological microequiprobability rule, were analyzed in the opening paragraphs of chapter 11.

Now suppose that you have no spear. You must decide whether to freeze or to run. Which is the better strategy? A complex question: either strategy might save you, and either might backfire. Many things ought to be taken into account; among them are the probability that the tiger sees you if you run and the probability that the tiger sees you if you freeze. These numbers depend on the tiger's orientation and the path of your retreat. Proportional dynamics and microequiprobability again come into play, as explained in section 9.3.

Also relevant—I assume that you remain as yet unseen—are the tiger's likely patterns of movement. In some circumstances the animal's wanderings may have the character of a random walk, as if, say, it hears leaves rustling and crisscrosses the area haphazardly looking for the source of the noise. Such a search can be modeled in much the same way that Darwin and his readers thought about the movement of pollinating insects, so that the probability of the predator's visiting a given area is proportional to the area's size (section 9.5). Even if the uniformity rule does not apply, probabilistic dynamics can play a role in your calculating better and worse places to hide, or in your estimating the time it will likely take for the tiger to happen upon you, as when you think about the time it will likely take for a ball in a shaken urn to enter the "target zone" (sections 6.4 and 8.2).

The more educated your guesses about these and other relevant probabilities, the more sophisticated your reasoning about the optimal strategy for tiger avoidance. Life is a game of chance. The better you can read those chances, the better you can play the odds.

One more example of the practical value of equidynamic expertise. You want to sneak into the shaman's hut. (Perhaps you are after her spear, perhaps her husband, perhaps something to soothe a tiger bite's sting.) Around the encampment, people come and go. What are the chances you will be detected in your attempt at infiltration? The trajectories of the people are, like the tiger's trajectory, part random and part nonrandom. Concerning the nonrandom part, equidynamics is no help, but with respect to the random part, it has something to say: for example, your chances of running into a passerby increase as the number of people on the move increases. Perhaps you would be better off making your move when everyone is at the campfire sing-along.

Obvious, no? Surely; why, though, is it so easy to see that the probability of detection increases (ceteris paribus) as the number of detectors increases? Attempt to quantify the relation between the two numbers based on your group's deterministic behavioral dynamics and you give yourself a monumental problem to solve, tracing the movements of every individual through the camp (section 9.2). To judge the reliability of a probabilistic approach to such problems is itself a deep and difficult task (Strevens 2003). What seems obvious to us, then, is practically impossible for a normal person to derive from scratch.

But normal people do not start from nothing. With equidynamics pre-installed in our heads, we can make quick and reliable judgments, often quantitative, about relevance in complex systems. Since we live in complex systems, and what happens in these systems determines when or whether we live, love, and die, it is well worth our while to develop this capacity to reason equidynamically. It would not be such a surprise, then, if it turns out that nature has generously made the investment on our behalf.

14.2 Equidynamics and Empiricism

Equidynamic reasoning, if what I have written in the preceding thirteen chapters is correct, has played and continues to play a major part in science. It has supplied the probabilistic framework, and a justification of that framework, for kinetic theory and so for statistical mechanics. It has done the same for evolutionary theory. And it continues to guide model-building in biology and elsewhere.

Equidynamics plays a major part in science, but it gets no credit. Scientists think equidynamically, but without, it seems, full awareness of their train of thought. They do not see clearly which aspects of a system—symmetries, smoothness, sensitivity—support their reasoning; nor do they see clearly what kind of reasoning it is. Simply to claim that reasoning from the nonstatistical physical properties of a system to physical probabilities is widespread and important in science is to say something controversial, or at least, unusual. Compare to the claim that any of the following figure prominently in scientific reasoning: repeatable controlled experimentation, observation, inference to the best explanation, mathematics—even thought experiments.

Why, for equidynamics, is there a relative lack of methodological self-consciousness? Consider three possible explanations.

First, equidynamic reasoning might be ignored because it is automatic and unconscious: physical premises flow in, and conclusions about physical probability flow out, of the equidynamic black box without our being in the least aware of the source of these excellent, intuitive, "obvious" judgments about the likelihoods of things.

The black-boxing of inference is familiar enough: perception, in particular, takes place behind a mental veil. We luxuriate in the end product of perceptual processing—an exquisite awareness of a world full of moving, interacting objects—while remaining oblivious to the multitudinous algorithms by which our brains assemble this rich mental picture from the raw material of the world's proddings of our nerve endings. If we want to understand how it all works, we have to reverse-engineer the box, just as though it were sitting in some other kind of organism's head.

Could it be this way with equidynamics too? We do find it hard to discern the workings of equidynamic reasoning without a considerable degree of reverse-engineering, the beginnings of which have been undertaken in this book. The sophisticated equidynamic inferences of young children further suggest, as remarked earlier in this chapter, a degree of innateness to the whole process that is characteristic of cognitively impenetrable black boxes (Fodor 1983).

In order to explain the lack of explicit methodological comment on equidynamics, however, we would have to be unconscious not only of its workings, but of its existence as well. The appeal to impenetrability and the analogy to perception fail to deliver what is needed. We may be unable to introspect how vision works, but we do not need to be brain scientists to know that we are seeing things. The opacity of the mental equidynamics box cannot in itself explain, then, why science appears not to have noticed the importance of equidynamic reasoning in the arguments of Maxwell, Darwin, Laplace, and all the die-tossers, wheel-spinners, and urn artists in this world.

A second possible explanation for equidynamics' failure to appear in the writings of scientists, philosophers of science, or historians, is that we are short on the mental resources required to conceptualize equidynamic reasoning.

It was not always so (the second explanation continues). In the days of classical probability—in Laplace's era—the tie between symmetry and probability was explicit, and probabilities inferred from symmetries were (if for obscure reasons) considered worldly enough to predict and explain frequencies.

As subjective and physical probability peeled apart, however, symmetry went one way and worldliness the other. The notion that frequencies could be inferred from symmetry-based probabilities came to seem insupportable. Probabilities could be derived from symmetries, yes, but not the sort of probabilities that would find their way into scientific theories such as statistical mechanics and evolutionary biology. So argued British empiricists such as Herschel's critic R. L. Ellis and John Venn (Ellis 1850; Venn 1866); Bertrand's paradox delivered the coup de grace. Even as the notion of physical probability became scientifically essential, then, the only known bridge between such probabilities and symmetries or other non-statistical properties was obliterated.

This story seems not quite sufficient, still, to explain why physicists, biologists, and philosophers have been so reluctant to remark on the manifest fact of their own inferences from nonprobabilistic dynamics to physical probabilities. After all, as I observed in section 3.3, the prime philosophical objections to the principle of indifference disappear if it is conceived as a rule for inferring physical probabilities from physical symmetries and other dynamically relevant properties. The division of classical probability into epistemic and physical probability should have been accompanied by a parallel division of the principle of indifference into distinct epistemic and physical rules, and thus the recognition of the existence of equidynamics. But something blocked the way.

That something, I suggest, is a tension between equidynamic thinking and an empiricist tenet central to scientific thought. The tenet is this: all confirmation must be founded in instance confirmation. To put it another way, to qualify as a legitimate piece of empirical evidence, a datum must be a positive or negative instance of the hypothesis upon which it bears—as a black raven is a positive instance, and a white raven a negative instance, of the hypothesis that all ravens are black.[1]

Hypotheses so evidenced may then support higher-level hypotheses by way of other inductive relations, such as inference to the best explanation, but entry-level hypotheses must be confirmed by their instances.

Think of higher-level support as instance confirmation "flowing up" to the theoretical level, so that even the most abstruse theory is ultimately rooted, epistemically, in instances of the phenomenological generalizations that it explains.

The scientific evidence for the inflationary universe theory, for example, consists among other things in observations of the cosmic microwave background. These constitute instances of, and so confirm, a hypothesis about the distribution of energy in the background. The distribution hypothesis then confirms the theory that (to our knowledge) best explains it. The first of these confirmation relations is an instantiation relation; empirical evidence in science, according to the instantialist tenet, must always be brought to bear by way of a relation of this sort.

That scientific inference conforms to the instantialist principle is a strong claim in need of considerable elaboration and a concerted defense. These I will not provide here. Better, I think, to stay focused on equidynamics and to show you, in a frankly speculative spirit, how scientific instantialism might account for scientists' inability to grasp the nature of their own inferences from dynamics to physical probability.

The empirical facts that constitute the premises of an equidynamic inference are, you will observe, in no sense instances of the hypothesis about physical probability that serves as its conclusion. A claim about the physical probability of obtaining heads on a tossed coin, for example, has as its instances the outcomes of coin tosses, not the symmetries of the coin. A standard empirical test of such a claim, then, would tabulate outcomes, considering the claim to be confirmed to the degree that the frequency of heads matched the probability of heads. The symmetries are not outcomes, and so may not, on an instantialist approach to scientific reasoning, count as empirical evidence for the probabilities. Nothing else, however, enters into an equidynamic inference. Thus, equidynamic reasoning is, by instantialist lights, not based on empirical evidence.[2]

That does not in itself put such thinking off limits to the scientist. Equidynamics is a form of what is often called "physical intuition," and such intuition is generally recognized to be a useful aid to theoretical innovation and experimental design. Intuition is not, however, supposed to play a role in empirical testing itself; it may make suggestions and drop hints, but it does not supply anything but the most prima facie, indeed private, reasons to believe (or "accept") a theory. You cannot publish a hypothesis on the grounds that it is intuitive, only on the grounds that it predicts or explains the evidence.

But of course these rules are not always followed, and they are perhaps ignored most at science's most formative moments: in Maxwell's 1859 paper on kinetic theory (even if Maxwell is rather coy about his reasoning), or in Darwin's "long argument" of the same year for the plausibility of natural selection as an agent of radical biological change. Genius often consists largely in the creative violation of prevailing norms. "How extremely stupid not to have thought of that!"—or perhaps, how extremely orthodox.

Scientists, I suggest, would nevertheless prefer not to notice that crucial aspects of their theoretical edifice are, or were for decades, supported by a pillar of fact that rests on something nonevidential. They go along with equidynamic reasoning because equidynamic reasoning is intuitive, but they do so furtively and in bad faith. They turn their inner mental eye away from their epistemic sins even as they indulge. They have a bad conscience.

The consequences of repression can be seen quite plainly in the development of the major probabilistic theories of modern science (quantum mechanics, perhaps, aside). The eponymous probabilities of statistical mechanics—where do they come from? Please do not ask, is the physicists' official response; the theory works, and that is good enough (Tolman 1938, 70). From this point of view, the move from Boltzmann's dynamic conception of the statistical mechanical probabilities to Gibbs's static, formalist approach has been a great relief to physicists' epistemic modesty.[3] The rhetorical force of the canonical probability distributions of statistical mechanics comes to be based in institutional instinct—a socially approved method for constructing stochastic theories—rather than in the ecological validity of equidynamic rules of inference. The probabilities are torn from their bedding in human psychology; they are bled dry, petrified, made pure mathematical form. Predictions are nevertheless correctly made; the theory functions as intended. Hats off; a victory lap. Just do not ask why.

In evolutionary theory and population dynamics, the same ideology prevails. Probabilities are left out of the story whenever possible, as in the deterministic models of population ecology.[4] Where they are essential, they are presented as far as is possible as a part of the formal modeling apparatus, as in population genetics, rather than as a straightforward representation of nature's stochasticity.

A corollary: those few cases where the probabilistic facts are inferred from their instances, that is, from matching frequencies—those few cases where fitness is directly measured, such as Kettlewell's (1955) study of melanism in moths or the Grants' study of short-term evolution in Galápagos finches (Grant 1986)—are, for their sheer, rare, empirical legitimacy celebrated to a degree that goes far beyond their contribution to our understanding of evolutionary change.

14.3 How To Understand the History of Science

There are many ways to understand the history of science. This book takes what I believe is a largely unexplored route. To finish, a few words contrasting my approach to the history of probabilistic thinking in science with a previous and far better-known approach exemplified by Ian Hacking's *The Emergence of Probability* (Hacking 1975).

To explain why the concept of probability appeared in the West around 1660 and not before is Hacking's aim. His story is, you might say, epigenetic. We humans, and in particular we in the West, begin without a concept of probability. Cultural, intellectual, economic, and political forces, such as the fitful attentions of aristocratic gamblers, an interest in the foundations of legal reasoning, and an increasing demand for annuities, come together to coax the concept into being out of the ooze of mental prime matter. This new kind of mental jelly is given a backbone by philosophers and mathematicians such as Leibniz and Jacob Bernoulli. The era of classical probability has begun.

My story is by contrast preformationist. The concept of physical probability existed long before 1660. Or at least, something existed that was close enough to the probability concept to enable equidynamic thinking in both everyday life and scientific inquiry: something that could function as a predictive and explanatory bridge between dynamical properties and frequencies; something that had roughly the mathematical structure of frequencies (without which the predictive bridge could not be crossed); something that was applicable to the single case; something accompanied by the notion, or a near-equivalent, of stochastic independence.

The argument for the preexistence of the notion of probability is simple: without such a concept, equidynamic reasoning in six-month-old children and other unschooled types would be impossible. Infant equidynamics is demonstrably real. Therefore a concept having the enumerated properties is always already present in the human mind (section

4.4; as noted in that section, Franklin [2001] is a preformationist fellow traveler).

The story of "the emergence of probability," given this preformationist tenet, is a story not of the construction of a concept but of an emerging awareness of the nature of a concept that has been present all along. Probability appears not from nothing, but from inside the magic black hats of our own minds. The historical forces identified by Hacking should be understood, then, not so much as putting new ideas into the mental inventory (though the preformationist will allow a certain degree of conceptual evolution), as helping us to see what is already present, not least by removing barriers to and prejudices against certain styles of thought. Indeed, Hacking's narrative works rather well when recast along these lines: a widespread preference for nomological determinism can just as easily explain the suppression of explicitly statistical models of the world as it can explain the failure of the concept of physical probability to arise at all.

An epigeneticist looks largely to what is outside the mind, not because the mind is unimportant, but because the mental exterior is where most of the explanatory story lies. What is outside—cultural and social spurs or impediments to the expression of latent inferential tendencies—is important to the preformationist as well. But more important still is the internal constitution of the mind. A preformationist must be a psychologist or a cognitive scientist, as well as a philosopher or a historian.

Maxwell's derivation might be read by an epigeneticist as an expression of the particular social and intellectual forces of the time, demanding or at least making possible a certain historically new response to old scientific problems (or perhaps raising new questions altogether, such as—how are the velocities distributed?).

A preformationist does not doubt the existence or importance of the historical context, but they might nevertheless—as I have in sections 2.3 and 8.4—read the derivation as an expression of an older way of thinking, partly obscured by a superstructure of fleeting historical significance, in this case, Maxwell's homage to Herschel.

As my study of Maxwell shows, the preformationist historian is a close reader of arguments who focuses not so much on their overt structure as on what is missing from that structure—on the things that seem so obvious or plausible that they are allowed to pass without comment. What is transparent to scientists is often enough invisible to historians of science, and so the preformationist may have to argue strenuously for the importance or even the existence of their enthymematic windows into the mind.

Here you have the argument. Important parts of the history of statistical mechanics, evolutionary biology, and statistics itself are to be explained in part, I have contended, by appeal to equidynamics, and thus to the invocation under social and cultural stars correctly aligned of ancient powers of probabilistic reasoning.

How much of the rest of science's development can and should be understood from the preformationist stance? There is glamour to the epigeneticist's notion that great leaps forward in science are conceptual leaps, that they require the synthesis of entirely new ways of thinking. Further, new concepts have without question emerged in the course of scientifically revolutionary thought: inertia, the space-time interval, perhaps entropy. But for what proportion of the sum total of scientific progress is this true? How often does preformationism come closer to truth? How often is new science made, not through new forms of thought, but through old forms applied to new subject matter? Some of the oldest forms have perhaps taken scientific innovation furthest: spatialization, causality, probability—probability and equidynamics.

NOTES

GLOSSARY

REFERENCES

INDEX

NOTES

1. The Apriorist

1. Maxwell proposed a frequency distribution rather than a probability distribution; the distinction between probabilities and frequencies will however be elided in these early pages.

2. Likewise, another assumption made by earlier kinetic theorists, that molecules are equally likely to be traveling in any direction, seems to be borne out by the general stillness of the air in an enclosed space.

2. The Historical Way

1. On the reception of the kinetic theory, see Brush (1976, section 5.4). Maxwell had his doubts early as well as late: he thought that the result on the density-independence of viscosity was surely wrong, and he found that the theory predicted the wrong ratio of specific heats. The intuitively jarring prediction about viscosity turned out to be correct, redoubling Maxwell's interest in the theory, but he was never able to resolve the problem of specific heat. (Quantum mechanics eventually supplied the answer [Brush 1983, 148].) The kinetic theory continued to experience much opposition through the 1860s; its early supporters, however, included Hermann von Helmholtz and William Thomson (Lord Kelvin), both of whom appreciated Maxwell's contribution in particular.

2. In the case that Maxwell considers, which might be a gas in a box on the earth's surface, a gravitational field is of course present, but the difference in gravitational potential between the box's top and bottom is negligible.

3. In the original text, Herschel's justification for the assumption of Cartesian component independence is unclear, but when the review was reprinted several years later in 1857, Herschel added a footnote grounding the assumption in the observation that "the increase or diminution in one [component] may take place

without increasing or diminishing the other"; this is essentially Maxwell's reasoning (Herschel 1857, 399).

4. Herschel would have been truer to his earlier subjectivist rationale had he written "supposing no *knowledge* of bias, etc." I will suggest in section 11.1 that this lapse exposes the role of unvoiced physical considerations in guiding Herschel's statistical reasoning.

5. Garber et al. (1986, 9–11) provide a short and judicious overview of the possible influence of Herschel on Maxwell. Porter (1986) gives a fascinating treatment of the possible origins of Maxwell's statistical attitude to physics in the "social physics" of Quetelet and others.

6. This is true only for three-dimensional spheres, not for spheres of two or four dimensions (that is, circles or hyperspheres), or for nonspherical molecules.

7. The importance of the first three propositions has more often than not been overlooked in the secondary literature: Brush (1976, 186) notes it in passing (but clearly thinks that the propositions are utterly inadequate to their assigned task); Truesdell (1975) not at all. An unpublished paper by Balázs Gyenis, written independently of this book, is as far as I know the only substantial discussion— but it is unfortunately not currently available.

8. I do not claim that the argument for short-term instability is rigorous, merely that it has some intuitive force. The same goes for the corresponding argument for short-term stability in the next paragraph's reasoning. The source of such intuitions will be explored in section 7.2.

3. The Logical Way

1. Keynes (1921, chap. 4), who gave the principle of indifference its modern name, provides an excellent survey, as well as making of course his own contributions.

2. Jaynes' explanation for this transmutation of noninformation into knowledge (Jaynes 2003, 393–396) is worth your time even if it is at bottom unsatisfyingly equivocal. I cannot do justice here to the considerable literature on the Bertrand paradox and the often ingenious attempts to coax the principle of indifference into speaking unambiguously; for a recent response to Jaynes's and Marinoff's proposals, see Shackel (2007).

3. For the case against the possibility of such choices, however, see White (2005).

4. Physical and epistemic probability may be further subdivided. Physical probability appears to come in two varieties, one irreducible (Giere 1973) and one grounded in nonprobabilistic physical facts (Strevens 2011). Epistemic probability, as implied in the main text, comes in at least two varieties: subjective probability, representing an aspect of a rational agent's psychology, the raw

material of Bayesian epistemology (Howson and Urbach 1993), and what is often called logical probability, representing the inductive support that one set of propositions gives another (Keynes 1921). For an overview and the modern picture of how it all fits together, see Strevens (2006).

5. Williamson (2010, 152) allows, however, that in some circumstances there may be several equally good distributions.

6. It would be wrong, however, to suppose that Laplace's equal possibilities are our equal physical probabilities. Because Laplace does not distinguish physical and epistemic probability, his use of the term *equal possibility* is equivocal, and frequently inconsistent with the physical interpretation.

4. The Cognitive Way

1. Formally, a Bernoulli distribution is simply a probability distribution with two possible outcomes, such as "red" and "white" or "heads" and "tails." I use the term in a broader sense in what follows. First, I assume that outcomes from repeated trials are stochastically independent (technically, they constitute a "Bernoulli process"). Second, in later chapters, I allow that such a distribution may have more than two outcomes. So in chapter 6, I will say that the physical probability distribution over the outcomes of a series of die rolls is a Bernoulli distribution, meaning to imply that the outcomes are independent, and despite the fact that there are six possible outcomes rather than two. I hope you will grant me philosophical license to use the term "Bernoulli" in this terminology-minimizing way.

2. Between 70 percent and 80 percent of choices were directed to this cup. This was true whether a cup was closer to the jar whose sample it contained or closer to the other jar.

3. Franklin (2001, 322–323), who notably challenges the historical consensus, suggests that the same is true for the probabilistic concepts implicated in inductive reasoning. Carey provides guidance on inferring innateness from early competence and on the shadings in the notion of innateness itself.

4. I would vote for an indirect application: from the physical probability, a single-case epistemic probability is derived—essentially, Reichenbach's (1949) view. For the alternative, on which there exist single-case physical probabilities, see Giere (1973). Either version is good enough for equidynamics; the difference is, in my view, of metaphysical interest only.

5. An intermediate view is possible: children, and perhaps many adults, operate with a classical notion of probability that does not distinguish between physical and epistemic probabilities. On this view, equidynamics delivers judgments about what are in fact physical probabilities, but consumers of these judgments need not make the distinction to put them fruitfully to use.

5. Stirring

1. To mention a few: Reichenbach (1949, 1956), Salmon (1970), Clark (1987), Loewer (2001), Strevens (2003), Glynn (2010), Abrams (2012).

2. I am also assuming that the *red*-producing set is measurable with respect to the initial speed distribution and that veridical physical probability distributions cannot disagree about the probability of an event.

3. That the one-half probability of *red* is somehow explained by the wheel of fortune's microconstancy and strike ratio for *red* of one-half, or by a related property, has been observed by many writers, often under the rubric "the method of arbitrary functions" (von Kries 1886; Poincaré 1896; Reichenbach 2008; Hopf 1934; Zhang 1990; Engel 1992; Strevens 2003). A short history of this approach to thinking about physical probability is supplied by von Plato (1983); I explain the differences between my own and others' versions of the approach in Strevens (2003, section 2.A) and from a more metaphysical point of view in Strevens (2011).

4. Proofs are given in Strevens (2003, section 3.B1), as is a more general discussion of the relation between microconstancy and independence. For a worked example, see Strevens (manuscript).

5. When I say that a process's production of an outcome is "sensitive to initial conditions" I mean nothing more than what those English words customarily convey: small differences in initial conditions will make a difference as to whether or not the outcome occurs. Some philosophical writers object to this use on the grounds that chaos theory has reserved the expression "sensitivity to initial conditions" for a special kind of sensitivity (roughly, exponential divergence in the limit). They should be ignored. Scientists, mathematicians, and philosophers may, if they wish, give natural language expressions proprietary definitions in the narrow technical contexts in which they work, but to attempt to impose those definitions on the general population is illiberal and obnoxious.

6. Or at least, this is one possible choice of stirring parameter; typically, there are many. In the case of a wheel with the standard symmetrical paint scheme, a shrewder choice of stirring parameter would be the position of the pointer within the currently indicated red-black pair, so that the parameter cycles as many times in a single revolution of the wheel as there are red-black pairs in the wheel's paint scheme.

7. To talk of either smoothness or a proportion presupposes what probability theorists call a measure on the smoothness parameter. The measure-relativity of these and other notions used in equidynamic reasoning is discussed in section 5.6, where it is resolved using the notion of a standard variable.

8. Philosophers of statistical mechanics will note that a stirring mechanism is ergodic, in the old sense, with respect to its stirring parameter. Equidynamics does not consider this ergodicity sufficient grounds to infer the existence of a

physical probability, however; the cycling's smoothness and multiplicity are required as well.

9. I assume that the coin is sent spinning perfectly around one of its diameters; in fact, such perfection is rare, and departures from perfection can make a difference to the probability of heads, albeit small. On this point and its history, and on complexities in the relation between physics and probability in the coin toss more generally, Diaconis et al. (2007) is an authoritative source.

10. In probability mathematics a random variable can take as its domain any set having an appropriate structure; it is not limited to physical quantities.

11. When philosophers talk about the possibility of measuring a physical quantity in two different ways, they often mean not using two different random variables, but using two different measures in the measure-theoretic sense over the space of the quantity's possible values. These are formally different ways of raising the same philosophical questions about the relativity of quantification. The measure in the philosophers' sense can be thought of as the composition of the random variable mapping the quantity's possible values to some set of numbers such as the reals, and a measure on those numbers (on the reals, usually the Lebesgue measure); see, for example, North (2010, 28).

6. Shaking

1. The tosses represented by the top half of the strip have gone on for slightly longer than those represented by the bottom half. In a standard toss, they will have gone on longer because the coin has been tossed higher. Consequently, in the top-half tosses the coin will have a greater downward translational velocity when it hits the floor than in the bottom-half tosses. I assumed above that this difference in velocities is not physically important; that is, that it is small enough that orientation alone makes almost all of the difference to the bouncing dynamics of top-half as opposed to bottom-half tosses.

The physical plausibility of this assumption tacitly depends, I think, on the floor's absorbing much of the impact of the landing, but then so does the randomizing power of the toss: if the landing were perfectly elastic, the bounce would send a coin spinning in precisely the reverse direction for about the same amount of time as the original toss, almost exactly undoing the stirring effect of the toss.

Note that strips on the extreme "top left" of the evolution function—that is, coins that are tossed to a great height, but spinning only very slowly—will be very tall; for these strips, the assumption that differences in landing speed between top-half and bottom-half tosses have little effect on the coin's bouncing is not so plausible. Two possible solutions: first, assume that such tosses are improbable (they might well be ruled illegitimate in a gambling context); or second, give the strips a different orientation—say, perpendicular to the curves separating gray and white—and assume that neither small differences in translational velocity nor small differences in angular velocity make a sizeable contribution, within a strip, to the outcome of a bounce.

A third solution, which would make the whole argument much simpler, assumes that the coin is equally likely to be facing heads-uppermost as tails-uppermost at the beginning of a toss; this supposition, however, puts an unacceptable restriction on the scope of the argument: we want to show (because humans are disposed to believe) that the bouncing coin has a one-half probability of landing heads regardless of its initial orientation.

2. I will assume that precession and other fancy contributions to the die's motion are negligible.

3. My source is the Wikipedia entry on "Dice Control," consulted on October 12, 2010. It seems that the army blanket roll may have involved more sliding than rolling; it also seems to have been a roll using two dice rather than a single die, which makes outcomes rather harder to control.

4. The casino die roll—which according to the regulations need involve no shaking at all, and even allows gamblers to "set" the dice, or in other words to arrange them in a certain orientation before throwing—is much easier to analyze than the die roll with preliminary shaking, because the physics of a single bounce off the alligator wall is relatively tractable.

5. Further experimental philosophy suggests that another kind of at-home die roll, tossing the die onto a hard surface from a polite height, is remarkably nonrandom in its dynamics. The fairness of such rolls is presumably due in great part to the roller's casual and inattentive handling of the dice before tossing, which is somewhat like shaking but less reliable in its randomizing effects.

6. Readers who know enough to care may add to these claims the necessary qualifications about measurability and so on.

7. I am of course ignoring the slight physical differences that realize the die's face-numbering scheme.

8. Even with such an asymmetry among the quadrants, the evolution function will tend to microconstancy as the level of embedding increases: almost all regions of the initial conditions will, given sufficiently many levels of embedding, for purely combinatorial reasons undergo roughly equal numbers of the four different asymmetric transformations effected by the inscription schemes in the four quadrants. A relatively short shaking of a die will not achieve a high-enough level of embedding for this effect to come into play, I think, but it will be relevant to other shaking processes, including perhaps the shaking of balls in an urn.

9. That was rather quick; there is of course quite a bit more to say about the conditions under which the dependence of later collisions' initial conditions on earlier initial conditions should be "much the same" whatever the approximate value of the earlier conditions. Since my topic is not primarily the physics of die

rolls, the complications and qualifications are omitted here; I am content to make a prima facie case that the shaking process has a microconstant dynamics.

10. On my use of the term *Bernoulli* for probability distributions with more than two outcomes, see note 1 of chapter 4.

11. Note the incipient correspondences to some important mathematical notions: the process of embedding, if taken to the limit, tends to create a fractal shape, while equal visitation rates are a kind of ergodicity, with the sort found in shaking processes bringing to mind modern ergodic theory and (as noted in note 8 of chapter 5) the sort found in stirring processes reminiscent of the original notion of ergodicity.

7. Bouncing

1. Independence of micropositions and such-like is, in the case where distributions over individual positions are microequiprobabilistic, entailed by the microequiprobability of the relevant joint density (Strevens 2003, section 3.43).

The three independence properties just described are entailed, then, by the microequiprobability of a joint density encompassing the pre- and post-interaction values of the state variables of all entities in a system. There is a pleasing unity to this way of framing things, but for expository purposes, in the main text I continue to identify the independence properties separately.

2. The straightforward proof is given in Strevens (2003); see proposition 3.5 (p. 219). A microequiprobabilistic marginal distribution for each state variable may also be inferred, but it is the conditional distribution that matters for the application of the equilibrium rule in the next section.

3. There is, you will see, a role for "relaxation" to play in the application of the microdynamic rule, since time strengthens microequiprobability; unlike the long-term rules investigated in the remainder of this chapter, however, the microdynamic rule tells you something about probability distributions in systems that are not in equilibrium.

4. Another complication is presented by Kolmogorov, Arnold, and Moser's KAM theorem, which shows that in some systems to which the microdynamic and equilibrium rules may be applied, the components of the system are not engaged in a true random walk. (More exactly, there are certain regions of state space of nonzero measure that a system can never leave: if a system starts in such a region, it remains there indefinitely. This implies that the motions of the system's parts are not fully stochastically independent.) The implications of the theorem for equidynamics are unclear. It is not certain that the systems to which we apply the microdynamic and equilibrium rules satisfy the conditions under which the KAM theorem applies. The dynamics of a system starting out in one of the regions in question might in any case be sufficiently similar to a true random walk that the predictions made by the microdynamic and equilibrium rules concerning a

system with such a starting point are qualitatively correct. And the regions in question might be so small that the chance of a system starting out in one is negligible (Sklar 1993, 169–174). The KAM theorem need not, then, stand in the way of the ergodic justification.

5. This is an extrapolation; Téglás et al.'s data does not establish a precise quantitative relationship between the probabilities.

8. Unifying

1. You can presumably ignore the possibility that the bouncing exactly reverses the effect of the shaking (see note 1 of chapter 6).

2. In the jargon, frequencies are "sufficient statistics" for Bernoulli processes.

3. The equiprobability of direction given magnitude is a stronger property than the independence of direction and magnitude. But given the equiprobability of direction, the two are equivalent. The reconstruction of the unofficial argument in section 2.3 turns on the short-term instability of the stronger property. An alternative derivation that is perhaps easier for equidynamicists to grasp and so more persuasive would have as its premises the short-term instability of the weaker property along with the equiprobability of direction, the latter premise being supplied directly by equidynamics' uniformity rule.

Another possibility is that there exists a stronger version of the uniformity rule than the one stated in chapter 7, a version that guarantees not only that the state variable in question is uniformly distributed at equilibrium, but also that its distribution conditional on the values of all other state variables is uniform. Such a rule would, if applicable to direction of travel, secure the equiprobability of direction given magnitude directly, which would provide an even stronger explanation of the unequal intuitive plausibility of the two Herschelian posits.

4. Proposition VI deals with equipartition among velocities of molecules with different masses; proposition XXIII covers equipartition among translational and rotational movement.

9. 1859 Again

1. I take some license with the dates. Maxwell's velocity distribution was unveiled in an 1859 talk to the British Association for the Advancement of Science and subsequently published in 1860. Darwin's *Origin* was published in 1859, but its arguments had been briefly presented (alongside Alfred Russel Wallace's ideas) in an 1858 paper read to the Linnean Society. That said, the particular passages from the *Origin* analyzed in this chapter were introduced to the world, like Maxwell's first four propositions, in 1859.

2. Various qualifications and refinements to this scheme are possible and surely desirable; Godfrey-Smith (2009) surveys and assesses the options, and more gen-

erally notes the limitations of attempts to provide an abstract characterization of the conditions under which natural selection occurs. Note that throughout this chapter I use the term *fitness* in a loose and broad sense, ignoring for simplicity's sake the distinction that many have drawn between a concept that is more closely tied to Darwin's notion of adaptive advantage and a concept that is allied with the mathematical machinery of modern Darwinism.

3. An almost identical case introduces natural selection in Darwin's 1842 sketch and his 1844 essay, the first versions of the theory in written form (Darwin 1909, 8–9, 91–92).

4. Hodge (1987) notes that Darwin's talk of the fitter individuals' "greater chances of success" begins in his notebooks of the late 1830s.

5. He writes: "For instance a vast number of eggs or seeds are annually devoured . . . Yet many of these eggs or seeds would perhaps, if not destroyed, have yielded individuals better adapted to their conditions of life than any of those which happened to survive" (Darwin 1872, 68).

6. See Sheynin (1980, 356, note 60) for this verdict along with minor exceptions.

7. Alfred Russel Wallace, in the independent exposition of evolution by natural selection that was read to the Linnean Society along with Darwin's abstract of the *Origin* in 1858, invokes the "law of averages" to account for the near-certainty with which a fitter variant will come to dominate its population (Wallace 1871, 37–38).

8. One exception: Gigerenzer et al. (1989, 66) write that Darwin "scarcely even made use of what we may call statistical reasoning," treating the natural selection of the fitter variant as a sure thing, as in artificial selection. This view seems to me to have difficulty explaining Darwin's liberal use of stochastic language.

9. Let me take a moment to dismiss a possible distraction. There are cases you can imagine in which the swiftest wolves will not be the best hunters of deer. Suppose, for example, that the swifter a wolf, the higher its metabolism, and that wolves with higher metabolisms can be more easily detected from a distance by deer's infrared detectors. Then it might be that at a certain point, further swiftness is counterproductive: running faster will not help you catch deer that can see you coming from miles away. Darwin is, of course, implicitly putting all such difficulties aside. His claim to be providing an "imaginary illustration" only is presumably made not because deer and wolves are imaginary (indeed, he reports a paragraph later on the apparent evolution of an especially swift wolf specializing in deer in the Catskill Mountains), but because he wants the reader to imagine a case in which there are no hidden costs to increased swiftness or other complications. Go ahead and explain these additional, unspoken qualifications to the Laplacean predictors; the question about the relative benefits of speed and fur will still seem to them to be grossly underspecified.

10. I assume that the initial conditions are sufficiently coarse-grained that they do not determine the prey's strategy, but also sufficiently fine-grained that they determine what will happen conditional on the strategy. In a case such as this, the assumption seems reasonable because the proximal strategy-determining conditions are internal to the prey (neural signals and so on), while the success-determining conditions are largely external.

11. In general, we will weight the sets by degree of help and hurt; in the case in question, the degrees are equal, so the question of weighting may be silently passed over.

12. I include in this condition of microequiprobability the sort of independence properties that were bundled into the notion of microequiprobability when it was first introduced at the beginning of chapter 3 and that are spelled out in the articulation of the microdynamic rule in section 7.1.

Observe that the microequiprobability condition allows that there may be several possible initial condition distributions, with the applicable distribution depending on the kind of encounter; for example, one distribution might apply during the day and one might apply at night (day and night being not themselves probabilistically distributed, you might reasonably suppose). What is assumed is that all such distributions are microequiprobabilistic.

13. The argument stated in the main text does not go through in every case. Suppose that "disadvantageous" initial conditions, though present throughout the initial condition space as sensitivity demands, are concentrated in certain macrosized regions, so that the ratio of disadvantageous to advantageous initial conditions is much higher in these regions than elsewhere. If the regions are also overwhelmingly favored by the actual initial condition distribution (which is consistent with the distribution's microequiprobability), then although advantageous initial conditions may predominate overall, the factor in question might be a probability lowerer. Exercise to the reader: give an informal argument that such cases are rare, from which it follows that the main text's reasoning is ecologically valid, that is, that it works in the great majority of actual cases.

14. Strevens (2003, sections 2.7, 3.B) supplies a general schema for partitioning information into high and low levels in this way.

15. Pollinating insects, like many other animals, "probability match" when foraging: given the choice between a richer and a poorer resource, they will choose one over the other with a frequency roughly proportional to the ratio of the resource qualities, rather than invariably choosing the richer resource (Greggers and Menzel 1993). If, for example, one resource is twice as rich as the other, they will choose the richer resource with a probability of two-thirds and the other with a probability of one-third. This behavior leads to a swarm of insects distributing themselves optimally from the point of view of the hive mind. But (I presume) neither Darwin nor his readers knew about probability matching (or the more complex behaviors described by Greggers and Menzel), and the character-

ization of the inference in the main text does not assume otherwise. Also of interest in this respect, but also presumably historically irrelevant, is the recent literature showing that some animals' foraging patterns can be modeled successfully as pure random walks rather than as nonrandom trajectories perturbed by the occasional random disturbance (Viswanathan et al. 2011).

10. Applied Bioequidynamics

1. There are, as we now know, complications: not only the number of expected offspring but other features of a variant's probabilistic reproductive tendencies, such as the variance in offspring number, make a difference to the speed and even the direction of selection (Gillespie 1977). Such subtleties presumably did not, and did not need to, play a role in nineteenth-century arguments for the importance of natural selection.

2. Darwin appeals to artificial selection to show that great physiological changes can come about from the concatenation of small selective steps, but natural and artificial selection differ precisely on the point in question: with artificial selection, there is no puzzle as to how the same traits can be selected over and over even as the surroundings change: there is a selector who, independently of the natural environment, chooses just such types.

3. On the suppression of probabilities, see section 14.2.

4. Equidynamics is equally useful for constructing models of neutral evolution; in that case, of course, its contribution is solely to declare that various properties are probabilistically irrelevant to selectively significant outcomes.

5. This is true regardless of the form of the resulting model. Thus even in cases where the model is deterministic, as when drift is "idealized away" by assuming infinitely large populations, stochastic thinking may legitimately be used to determine which aspects of the underlying ecological processes are relevant to selection.

6. Such variables might appear in an agent-based model, though I would not consider such a model a proper explanation of evolutionary change, containing as it does a surfeit of explanatorily irrelevant detail (Strevens 2008, section 8.33).

11. Inaccuracy, Error, and Other Fluctuations

1. My historical narrative relies principally on Stigler (1986), with some triangulation from Hald (1998) and my own reading of the primary sources.

2. Galileo suggested that the best estimate minimizes the sum of the differences between the estimate and the observed value; this amounts, though Galileo may not have realized it, to the method of taking the median rather than the mean—a different notion of "average" than that discussed in the main text. Galileo's

method, along with its vaguenesses and imprecisions, is discussed by Hald (1986).

3. I will write as though these are assumptions about a physical distribution, either probabilistic or frequentist, but the practice of averaging might equally well be justified by symmetry and independence in what we would now call an epistemic probability distribution, a possibility that will be considered in due course.

4. The formula is a simplified version of that used by Tobias Mayer to measure the moon's libration, as described below.

5. Mayer was working with a set of twenty-seven observations, but as in the example was solving for three variables.

6. Gauss later claimed to have invented the method and to have used it privately ten years before Legendre, spurring a priority dispute.

7. Hald (1998, section 6.9) provides a sympathetic though speculative reconstruction of Legendre's motivation for endorsing the least squares method.

8. Although for expository purposes the figure shows Simpson's distribution as continuous, it was in fact discrete.

9. Like Bayes' other, better-known work on probability, the critique went unpublished, though it must have been circulated, as Simpson learned of it soon after.

10. Gauss was possibly inspired by J. H. Lambert, who like Simpson and Laplace assumed a symmetrical error curve in which errors of greater magnitude have lower probability (Sheynin 1966).

11. Stigler (1986, 201–202) attributes the hypothesis of elementary errors to Laplace's 1810 paper, I think incorrectly. The hypothesis does appear in an embryonic form in the work of Daniel Bernoulli in 1778; like everyone else, Bernoulli assumes a symmetric error distribution (Hald 1998, 84).

12. Is it not obvious to the discerning observer that the fluctuations causing a star's twinkling are symmetrically distributed? No; it is unlikely, I think, that you would be able to tell the difference between twinkling that is symmetrical and twinkling that favors the top left over the bottom right quadrant by a ratio of 4:3. Some further reason must underlie your assumption that the twinkling you see is symmetrical.

13. Given the topic of this book, it is not entirely irrelevant to mention, if only in passing, Sheynin's (1984) overview of early statistical approaches to understanding the weather.

14. I rely for the history of mechanical forecasting principally on Edwards (2010).

15. What about chaos? That the weather is highly sensitive to initial conditions was already well known, or at least strongly suspected, when Phillips built his model. This sensitivity was and is thought to manifest itself in the medium term on the medium scale, for example, in the formation of fronts ten days out. It does not (as we know now) seriously impede short-term forecast-

ing; nor (as Phillips' contemporaries would have presumed) does it undermine large-scale structures. A small disturbance might, then, make a difference to whether or not a cyclone forms in a particular place two weeks later, but it does not make a difference to the fact that the mid-latitudes are a zone of frequent cyclone formation, nor to any of the other structural facts that Phillips hoped to explain.

16. This claim is based on a gut-level assessment of what "micro" amounts to; as I will observe in section 12.3, the standard for "micro" when applying the microequiprobability rule may be a complex and contextual matter.

17. Some macrolevel models—including those of kinetic theory and statistical physics, evolutionary biology, and population ecology—suppose something stronger, I argue in Strevens (2003), namely, that microlevel fluctuations cancel out, that is, that their expected contribution to macrolevel parameters is not merely limited, but zero.

18. Tolstoy, *War and Peace*, in Pevear and Volokhonsky's translation (Tolstoy 2007, 1021).

12. The Exogenous Zone

1. Cirripedologists may need to substitute a more appropriate example.

2. In what follows, as elsewhere, I assume that a quantity's being microequiprobabilistically distributed implies that the joint distribution over "sufficiently separated" multiple instances of that quantity is microequiprobabilistic. The microequiprobability of coin toss spin speeds, for example, implies that the distribution over the n speeds of any n coins tossed simultaneously by different croupiers is microequiprobabilistic, as is the distribution over the m speeds of any m coins tossed in sequence by the same croupier, and so on.

3. In what follows I use "variable" to mean "random variable." The nature of random variables is explained in section 5.6.

4. I assume that the oscillation in v^*'s density is microsized; the definition of v^* can of course be tuned to suit.

5. North (2010) offers a rather different approach to formulating a rule of this sort, on which a uniform distribution, rather than a merely microequiprobabilistic distribution, can be assumed, though over a much smaller set of variables, namely, the "canonical variables" of fundamental physics.

6. I am supposing that the magnitude of the noise is enough to impose uniformity on the relevant "micro" scale; the physiological facts support this assumption for the meaning of "micro" relevant to typical gambling devices. Note, by the way, that there is a certain minimum amount of noise that is present regardless of the size of the action. Thus a minuscule application of muscular force does not have a minuscule degree of variability; the variability bottoms out at some point, decreasing no further as the muscular force goes to zero.

7. What follows is a version of the "perturbation explanation" of microequiprobability presented in Strevens (2003, section 2.53, where it is called the "perturbation argument").

8. Why do I supply (1) and (2) as well as (3), when the conclusion follows from (3) alone? Because I am giving you not only an argument but an explanation, and (1) and (2) are what explain (3).

9. As mentioned in note 2 of chapter 7, Strevens (2003, 219) shows that if the joint distribution over a set of random variables is microequiprobabilistic, then any conditional distribution is also microequiprobabilistic. The distribution over the outcome conditions is the weighted sum of these conditional distributions.

10. The iterated explanation is close to the explanation tendered in Strevens (2003). Compare also the spirit of Poincaré's (1914, 83–86) remarks on the universal tendency to uniformity attributable to the second law of thermodynamics.

11. The simplest such process would produce the same output regardless of the input.

12. If you have Latin rather than Teutonic tastes in neologism, you might prefer "quasiprobability"; that term was used to pick out a different thing, however, in Strevens (2008).

13. And the relevant joint distributions over sets of rolls would also likely still have been smooth—this tendency providing an explanation for those more subtle kinds of frequentistic facts that turn on stochastic independence among the outcomes, such as the frequency of *red* in a series of rolls being the same whether or not the immediately preceding outcome was *red*.

14. I ignore those rather contrived circumstances under which a set of initial conditions might have a tendency to smoothness in the first sense without in actuality falling under a physical probability distribution.

15. If taking this route, something to put on the "to do" list: frame the perturbation explanation so that it does not assume a physical probability distribution over ur-conditions or outcome conditions, talking rather entirely in terms of distributions of frequencies.

13. The Elements of Equidynamics

1. On this heroic period in experimental stochastics see also Czuber's (1908, 149–151) report on the Zurich astronomer Rudolf Wolf's prodigious feats of die-rolling. (Fortunately, Labby [2009] has now automated the process.) Czuber concludes that Wolf's dice were, like Weldon's, not quite symmetrical. A perfect cube is hard to find, but equidynamics endures.

2. On the connections Leibniz makes between possibility, facility, and probability, see Hacking (1975, chap. 14).

14. Prehistory and Meta-History

1. I have in mind a more generous conception of instancehood than Hempel's (1945) version; background knowledge and auxiliary hypotheses may play a role in determining that something is an instance, as for example in Glymour (1980).

2. Instantialism does not militate against a scientific argument for a probability distribution that deduces the distribution from other probability distributions that are themselves confirmed statistically, that is, by their instances. In this sort of case, instance confirmation "flows down" the entailment relation.

3. The process of formalization perhaps begins at the same time as statistical physics itself, with Maxwell's decision—if my hypothesis in section 2.3 is correct—to paint over his original equidynamic argument for the proposition IV posits with the "official" a priori argument.

4. The reasons for omission are complex, I should add: there are legitimate reasons to explain stochastically generated phenomena with deterministic models (Strevens 2008, section 12.1).

GLOSSARY

The brief characterizations in this glossary are intended to jog your memory; they are not official or complete or formally rigorous definitions. I have typically not made explicit, for example, the need to apply equidynamic rules with respect to standard variables.

Equidynamics The system of rules we humans use to infer physical probabilities, or properties of physical probability distributions, from nonstatistical facts about the causal structure of the system that produces the outcomes in question. (3.3)

Equilibrium rule An equidynamic rule according to which you may infer that the probability distribution over a variable has reached a stable equilibrium, if the dynamics determining the variable's value satisfies conditions of randomization, boundedness, and relaxation. (7.2)

Equilibrium rule package The suite of equidynamic rules consisting of the microdynamic rule, the equilibrium rule, and the uniformity rule. (8.0)

Evolution function A function that represents the dynamics of a system with respect to a designated outcome by mapping initial conditions that produce the outcome to the number one and initial conditions that produce any other outcome to zero. (5.2)

Inflationary A dynamics is inflationary just in case it is sensitive to initial conditions, in the broadest sense of the phrase. (6.2)

Majority rule An equidynamic rule according to which you may infer that a causal factor increases the probability of an outcome if many more initial conditions lead to the outcome when the factor is present than when it is absent. Some variants of the rule are given in section 13.4. (9.3)

Microconstancy An evolution function for a designated outcome is microconstant if its domain—the space of initial conditions for the process it

245

represents—can be divided into many small (or "microsized") contiguous regions, in each of which the same proportion of initial conditions produce the outcome. That proportion is called the outcome's *strike ratio*. (5.2)

Microdynamic rule An equidynamic rule according to which you may infer that the probability distribution over a variable whose value is determined by iterated interactions is microequiprobabilistic, if the interaction dynamics satisfies conditions of smoothness and sensitivity to initial conditions. (7.1)

Microequiprobability A probability distribution is microequiprobabilistic if over any small contiguous interval or region, its density is approximately uniform. In this book, microequiprobability implies also that outcomes sampled from the distribution have a certain degree of stochastic independence (minimally, relevant joint distributions must be themselves microequiprobabilistic). (3.0; 5.2)

Microequiprobability rule An equidynamic rule according to which you may defeasibly assume that the probability distribution over any standard variable is microequiprobabilistic. (12.1)

Microlinearity A dynamical process is microlinear if its effect on any microsized, contiguous region of initial condition space can be approximated by a linear function. (7.1)

Microposition An entity's microposition is its position within a contiguous microsized area. By extension, its microdirection is its direction of travel within a contiguous microsized range of possible directions. (7.1)

Physiological microequiprobability rule A version of the microequiprobability rule according to which you may defeasibly assume that the probability distribution over quantities generated by the action of humans and other animals is microequiprobabilistic. It puts tighter constraints than the more general rule on what counts as "microsized," and so on what counts as microequiprobabilistic. (12.2)

Probabilistic dynamics A method for reasoning about a stochastic system that uses the system's underlying stochastic dynamics to estimate the probability that a given outcome occurs (or to infer some qualitative fact about such probabilities). It is used, for example, to estimate relaxation time when applying the equilibrium rule. (7.2)

Proportional dynamics A method for reasoning about a more or less deterministic system that uses the system's underlying deterministic dynamics to estimate the proportion of initial conditions leading to a given outcome (or to infer some qualitative fact about such proportions). It is used, for example, to determine that allowing a tossed coin to bounce does not affect the probability of heads. (6.1)

Shaking A shaking process is a randomizing process that over a certain period of time visits the values of a "shaking parameter" in such a way that every value has an "equal opportunity" of being visited (as when, for example, the process takes a random walk through the parameter space); the final value of the shaking parameter then determines the outcome. Examples of processes with a shaking component include the rolled die and sampling from an urn. Probabilities of outcomes of a shaking process are inferred using the equilibrium rule package, consisting of the microdynamic rule, the equilibrium rule, and often the uniformity rule. (6.3)

Standard variables The variables with respect to which the rules of equidynamics are defined. I take them to be, typically, proportional to the SI units. Their most important feature is their tendency to be or to become microequiprobabilistically distributed. (5.6)

Stirring A stirring process is a randomizing process that over a certain period of time cycles smoothly through all the values of a "stirring parameter"; the final value of the stirring parameter then determines the outcome. Examples of stirring processes include the tossed coin and the wheel of fortune. (5.4)

Stirring rule An equidynamic rule according to which you may infer a probability for an outcome produced by a stirring process equal to the outcome's strike ratio. You may also infer the stochastic independence of the outcomes of separate trials. (5.4)

Stochastic variable A variable responsible for randomizing an entity's trajectory through some set of possible states, invoked in the application of the equilibrium rule. (7.2)

Strike ratio Of a microconstant evolution function for a designated outcome, the proportion of initial conditions in each small region that produce the outcome. (5.2)

Uniformity rule An equidynamic rule according to which you may infer that the probability distribution over a variable is in the long term uniform, if the dynamics determining the variable's value satisfies the conditions required by the equilibrium rule and a further condition requiring a short-term tendency toward uniformity. (7.2)

REFERENCES

Abrams, M. (2012). Mechanistic probability. *Synthese* 187:343–375.

Bertrand, J. (1889). *Calcul des Probabilités*. Gauthier-Villars, Paris.

Bessel, F. W. (1838). Untersuchungen über die Wahrscheinlichkeit der Beobachtungsfehler. *Astronomische Nachrichten* 15:369–404.

Boltzmann, L. (1964). *Lectures on Gas Theory*. Translated by S. G. Brush. University of California Press, Berkeley, CA.

Browne, J. (1995). *Charles Darwin: Voyaging*. Jonathan Cape, London.

Brush, S. G. (1976). *The Kind of Motion We Call Heat: A History of the Kinetic Theory of Gases in the Nineteenth Century*. North-Holland, Amsterdam.

———. (1983). *Statistical Physics and the Atomic Theory of Matter from Boyle and Newton to Landau and Onsager*. Princeton University Press, Princeton, NJ.

Carey, S. (2009). *The Origin of Concepts*. Oxford University Press, Oxford.

Carnap, R. (1950). *Logical Foundations of Probability*. University of Chicago Press, Chicago.

Clark, P. (1987). Determinism and probability in physics. *Proceedings of the Aristotelian Society* 61:185–210.

Cournot, A. A. (1843). *Exposition de la Théorie des Chances et des Probabilités*. Hachette, Paris.

Croson, R. and J. Sundali. (2005). The gambler's fallacy and the hot hand: Empirical data from casinos. *Journal of Risk and Uncertainty* 30:195–209.

Czuber, E. (1908). *Wahrscheinlichkeitsrechnung*. Second edition. Teubner, Leipzig.

Darwin, C. (1859). *On the Origin of Species by Means of Natural Selection, or the Preservation of Favoured Races in the Struggle for Life*. First edition. John Murray, London.

———. (1872). *The Origin of Species by Means of Natural Selection, or the Preservation of Favoured Races in the Struggle for Life*. Sixth edition. John Murray, London.

————. (1909). *The Foundations of the Origin of Species*. Edited by F. Darwin. Cambridge University Press, Cambridge.

Daston, L. (1988). *Classical Probability in the Enlightenment*. Princeton University Press, Princeton, NJ.

Denison, S., C. Reed, and F. Xu. (2013). The emergence of probabilistic reasoning in very young infants: Evidence from 4.5- and 6-month-olds. *Developmental Psychology* 49:243–249.

Denison, S. and F. Xu. (2010a). Integrating physical constraints in statistical inference by 11-month-old infants. *Cognitive Science* 34:885–908.

————. (2010b). Twelve- to 14-month-old infants can predict single-event probability with large set sizes. *Developmental Science* 13:798–803.

Diaconis, P., S. Holmes, and R. Montgomery. (2007). Dynamical bias in the coin toss. *SIAM Review* 49:211–235.

Edwards, P. N. (2010). *A Vast Machine: Computer Models, Climate Data, and the Politics of Global Warming*. MIT Press, Cambridge, MA.

Ellis, R. L. (1850). Remarks on an alleged proof of the "Method of Least Squares," contained in a late number of the Edinburgh Review. *Philosophical Magazine* 37:321–328.

Endler, J. A. (1986). *Natural Selection in the Wild*. Princeton University Press, Princeton, NJ.

Engel, E. (1992). *A Road to Randomness in Physical Systems*. Lecture Notes in Statistics, vol. 71. Springer-Verlag, Heidelberg.

Feynman, R. P., M. A. Gottlieb, and R. Leighton. (2006). *Feynman's Tips on Physics: A Problem-Solving Supplement to the Feynman Lectures on Physics*. Pearson, Boston.

Fodor, J. A. (1983). *The Modularity of Mind*. MIT Press, Cambridge, MA.

Franklin, J. (2001). *The Science of Conjecture: Evidence and Probability before Pascal*. Johns Hopkins, Baltimore, MD.

Galilei, G. (1962). *Dialogue Concerning the Two Chief World Systems, Ptolemaic and Copernican*. Second edition. Translated by S. Drake. University of California Press, Berkeley, CA.

Garber, E. (1972). Aspects of the introduction of probability into physics. *Centaurus* 17:11–39.

Garber, E., S. G. Brush, and C. W. F. Everitt (eds.). (1986). *Maxwell on Molecules and Gases*. MIT Press, Cambridge, MA.

Gauss, C. F. (1809). *Theoria Motus Corporum Celestium*. Perthes and Besser, Hamburg.

Giere, R. N. (1973). Objective single-case probabilities and the foundation of statistics. In P. Suppes, L. Henkin, G. C. Moisil, and A. Joja (eds.), *Logic, Methodology and Philosophy of Science IV: Proceedings of the Fourth International Congress for Logic, Methodology and Philosophy of Science, Bucharest, 1971*. North Holland, Amsterdam.

Gigerenzer, G., Z. Swijtink, T. Porter, L. Daston, J. Beatty, and L. Krüger. (1989). *The Empire of Chance: How Probability Changed Science and Everyday Life*. Cambridge University Press, Cambridge.

Gillespie, J. H. (1977). Natural selection for variances in offspring numbers: A new evolutionary principle. *The American Naturalist* 111:1010–1014.

Glymour, C. (1980). *Theory and Evidence*. Princeton University Press, Princeton, NJ.

Glynn, L. (2010). Deterministic chance. *British Journal for the Philosophy of Science* 61:51–80.

Godfrey-Smith, P. (2009). *Darwinian Populations and Natural Selection*. Oxford University Press, Oxford.

Grant, P. R. (1986). *Ecology and Evolution of Darwin's Finches*. Princeton University Press, Princeton, NJ.

Greggers, U. and R. Menzel. (1993). Memory dynamics and foraging strategies of honeybees. *Behavioral Ecology and Sociobiology* 32:17–29.

Hacking, I. (1975). *The Emergence of Probability*. Cambridge University Press, Cambridge.

Hagen, G. (1837). *Grundzüge der Wahrscheinlichkeits-Rechnung*. Dümmler, Berlin.

Hald, A. (1986). Galileo's statistical analysis of astronomical observations. *International Statistical Review* 54:211–220.

———. (1998). *A History of Mathematical Statistics from 1750 to 1930*. Wiley, New York.

Hempel, C. G. (1945). Studies in the logic of confirmation. *Mind* 54:1–26, 97–121.

Herschel, J. F. W. (1850). Quetelet on probabilities. *Edinburgh Review* 92:1–57.

———. (1857). *Essays From the Edinburgh and Quarterly Reviews: With Addresses and Other Pieces*. Longman, Brown, Green, Longmans, & Roberts, London.

Hodge, M. J. S. (1987). Natural selection as a causal, empirical and probabilistic theory. In L. Krüger, G. Gigerenzer, and M. S. Morgan (eds.), *The Probabilistic Revolution*, vol. 2. MIT Press, Cambridge, MA.

Holland, J. H., K. J. Holyoak, R. E. Nisbett, and P. R. Thagard. (1986). *Induction: Processes of Inference, Learning, and Discovery*. MIT Press, Cambridge, MA.

Hopf, E. (1934). On causality, statistics and probability. *Journal of Mathematics and Physics* 13:51–102.

Howson, C. and P. Urbach. (1993). *Scientific Reasoning: The Bayesian Approach*. Second edition. Open Court, Chicago.

Jaynes, E. T. (2003). *Probability Theory: The Logic of Science*. Cambridge University Press, Cambridge.

Jeffreys, H. (1939). *Theory of Probability.* Oxford University Press, Oxford.

Johnson-Laird, P. N., P. Legrenzi, V. Girotti, M. S. Legrenzi, and J.-P. Caverni. (1999). Naive probability: A mental model theory of extensional reasoning. *Psychological Review* 106:62–88.

Kahneman, D., P. Slovic, and A. Tversky. (1982). *Judgment under Uncertainty: Heuristics and Biases.* Cambridge University Press, Cambridge.

Keegan, J. (1976). *The Face of Battle.* Jonathan Cape, London.

Keller, J. (1986). The probability of heads. *American Mathematical Monthly* 93:191–197.

Kettlewell, H. B. D. (1955). Selection experiments on industrial melanism in the Lepidoptera. *Heredity* 9:323–342.

Keynes, J. M. (1921). *A Treatise on Probability.* Macmillan, London.

Krüger, L. (1987). The slow rise of probabilism: Philosophical arguments in the nineteenth century. In L. Krüger, L. J. Daston, and M. Heidelberger (eds.), *The Probabilistic Revolution,* vol. 1. MIT Press, Cambridge, MA.

Labby, Z. (2009). Weldon's dice, automated. *Chance* 22:6–13.

Laplace, P. S. (1810). Supplement au mémoire sur les approximations des formules qui sont fonctions de très grands nombres et sur leur applications aux probabilités. *Mémoires de l'Académie des Sciences* 10:559–565. *Oeuvres Complets* 12, 349–353.

———. (1902). *A Philosophical Essay on Probabilities.* Translated by F. W. Truscott and F. L. Emory. Wiley, New York.

———. (1986). Memoir on the probability of the causes of events. *Statistical Science* 1:364–378. Originally published in 1774. *Oeuvres Complets* 8, 27–65.

Legendre, A. M. (1805). *Nouvelles Méthodes pour la Détermination des Orbites des Comètes.* Courcier, Paris.

Lewis, D. (1980). A subjectivist's guide to objective chance. In R. C. Jeffrey (ed.), *Studies in Inductive Logic and Probability,* vol. 2. University of California Press, Berkeley, CA.

Lewis, J. M. (2000). Clarifying the dynamics of the general circulation: Phillips's 1956 experiment. In D. A. Randall (ed.), *General Circulation Model Development: Past, Present, and Future.* Academic Press, San Diego.

Loewer, B. (2001). Determinism and chance. *Studies in History and Philosophy of Modern Physics* 32:609–620.

Lotka, A. J. (1956). *Elements of Mathematical Biology.* Second edition. Dover Publications, New York. First edition published in 1924 as *Elements of Physical Biology.*

Lynch, P. (2006). *The Emergence of Numerical Weather Prediction: Richardson's Dream.* Cambridge University Press, Cambridge.

Marinoff, L. (1994). A resolution of Bertrand's paradox. *Philosophy of Science* 61:1–24.

Maxwell, J. C. (1860). Illustrations of the dynamical theory of gases. *Philosophical Magazine* 19–20:19–32, 21–37. Reprinted in Niven (1890), vol. 1, pp. 377–409, and in Garber et al. (1986), pp. 286–318. Page references are to Niven; these references are reproduced by Garber et al.

———. (1867). On the dynamical theory of gases. *Philosophical Transactions of the Royal Society of London* 157:49–88. Reprinted in Niven (1890), vol. 2, pp. 26–78, and in Garber et al. (1986), pp. 420–472. Page references are to Niven; these references are reproduced by Garber et al.

Mehler, J. and E. Dupoux. (1994). *What Infants Know: The New Cognitive Science of Early Development.* Translated by P. Southgate. Blackwell, Oxford.

Niven, W. D. (ed.). (1890). *Scientific Papers of James Clerk Maxwell.* Cambridge University Press, Cambridge. Two volumes.

North, J. (2010). An empirical approach to symmetry and probability. *Studies in History and Philosophy of Modern Physics* 41:27–40.

Norton, J. D. (2008). Ignorance and indifference. *Philosophy of Science* 75:45–68.

Novack, G. (2010). A defense of the principle of indifference. *Journal of Philosophical Logic* 39:655–678.

Oster, G. F. and E. O. Wilson. (1978). *Caste and Ecology in the Social Insects.* Princeton University Press, Princeton, NJ.

Pearson, K. (1900). On the criterion that a given system of deviations from the probable in the case of a correlated system of variables is such that it can be reasonably supposed to have arisen from random sampling. *Philosophical Magazine* 5:157–175.

Phillips, N. A. (1956). The general circulation of the atmosphere: A numerical experiment. *Quarterly Journal of the Royal Meteorological Society* 82:123–164.

Plackett, R. L. (1958). The principle of the arithmetic mean. *Biometrika* 45: 130–135.

Poincaré, H. (1896). *Calcul des Probabilités.* First edition. Gauthier-Villars, Paris.

———. (1914). *Science and Method.* Translated by F. Maitland. T. Nelson, London.

Polanyi, M. (1961). Knowing and being. *Mind* 70:458–470.

Porter, T. M. (1986). *The Rise of Statistical Thinking 1820–1900.* Princeton University Press, Princeton, NJ.

Reichenbach, H. (1949). *The Theory of Probability.* University of California Press, Berkeley, CA.

———. (1956). *The Direction of Time.* University of California Press, Berkeley, CA.

———. (2008). *The Concept of Probability in the Mathematical Representation of Reality.* Translated and edited by F. Eberhardt and C. Glymour. Open

Court, Chicago. Reichenbach's doctoral dissertation, originally published in 1916.

Richardson, L. F. (1922). *Weather Prediction by Numerical Process*. Cambridge University Press, Cambridge.

———. (1939). *Generalized Foreign Politics: A Study in Group Psychology*. Cambridge University Press, Cambridge. Published as a monograph supplement to *The British Journal of Psychology*.

Rohrlich, F. (1996). The unreasonable effectiveness of physical intuition: Success while ignoring objections. *Foundations of Physics* 26:1617–1626.

Salmon, W. C. (1970). Statistical explanation. In R. G. Colodny (ed.), *The Nature and Function of Scientific Theories,* pp. 173–231. University of Pittsburgh Press, Pittsburgh. Reprinted in Salmon et al. (1971).

Salmon, W. C., R. Jeffrey, and J. Greeno. (1971). *Statistical Explanation and Statistical Relevance*. University of Pittsburgh Press, Pittsburgh.

Schmidt, R. A., H. Zelaznik, B. Hawkins, J. S. Frank, and J. T. Quinn, Jr. (1979). Motor-output variability: A theory for the accuracy of rapid motor acts. *Psychological Review* 86:415–451.

Shackel, N. (2007). Bertrand's paradox and the principle of indifference. *Philosophy of Science* 74:150–175.

Sheynin, O. B. (1966). Origin of the theory of errors. *Nature* 211:1003–1004.

———. (1980). On the history of the statistical method in biology. *Archive for History of Exact Sciences* 22:323–371.

———. (1984). On the history of the statistical method in meteorology. *Archive for History of the Exact Sciences* 31:53–95.

Simon, H. A. (1959). Review of Lotka. *Econometrica* 27:493–495.

Sklar, L. (1993). *Physics and Chance*. Cambridge University Press, Cambridge.

Sperber, D., D. Premack, and A. J. Premack (eds.). (1995). *Causal Cognition*. Oxford University Press, Oxford.

Stern, O. (1920a). Eine direkte Messung der thermischen Molekulargeschwindigkeit. *Zeitschrift für Physik* 2:49–56.

———. (1920b). Nachtrag zu meiner Arbeit: "Eine direkte Messung der thermischen Molekulargeschwindigkeit." *Zeitschrift für Physik* 3:417–421.

Stigler, S. M. (1986). *The History of Statistics: The Measurement of Uncertainty before 1900*. Harvard University Press, Cambridge, MA.

Strevens, M. (1998). Inferring probabilities from symmetries. *Noûs* 32: 231–246.

———. (2003). *Bigger than Chaos: Understanding Complexity through Probability*. Harvard University Press, Cambridge, MA.

———. (2006). Probability and chance. In D. M. Borchert (ed.), *Encyclopedia of Philosophy,* second edition. Macmillan Reference USA, Detroit.

———. (2008). *Depth: An Account of Scientific Explanation*. Harvard University Press, Cambridge, MA.

————. (2011). Probability out of determinism. In C. Beisbart and S. Hartmann (eds.), *Probabilities In Physics*. Oxford University Press, Oxford.

————. (forthcoming). Stochastic independence and causal connection. *Erkenntnis*.

Téglás, E., V. Girotto, M. Gonzalez, and L. L. Bonatti. (2007). Intuitions of probabilities shape expectations about the future at 12 months and beyond. *Proceedings of the National Academy of Sciences* 104:19,156–19,159. Stimuli available online at http://www.pnas.org/content/suppl/2007/11/13/0700271104.DC1.

Téglás, E., E. Vul, V. Girotto, M. Gonzalez, J. B. Tenenbaum, and L. L. Bonatti. (2011). Pure reasoning in 12-month-old infants as probabilistic inference. *Science* 332:1054–1059.

Todorov, E. (2002). Cosine tuning minimizes motor errors. *Neural Computation* 14:1233–1260.

Tolman, R. C. (1938). *The Principles of Statistical Mechanics*. Oxford University Press, Oxford.

Tolstoy, L. (2007). *War and Peace*. Translated by R. Pevear and L. Volokhonsky. Knopf, New York.

Truesdell, C. (1975). Early kinetic theories of gases. *Archive for History of Exact Sciences* 15:1–66.

van Fraassen, B. C. (1989). *Laws and Symmetry*. Oxford University Press, Oxford.

Venn, J. (1866). *The Logic of Chance*. Macmillan, London.

Viswanathan, G. M., M. D. E. da Luz, E. P. Raposo, and H. E. Stanley. (2011). *The Physics of Foraging: An Introduction to Random Searches and Biological Encounters*. Cambridge University Press, Cambridge.

Volterra, V. (1926). Fluctuations in the abundance of a species considered mathematically. *Nature* 118:558–560.

von Kries, J. (1886). *Die Principien der Wahrscheinlichkeitsrechnung*. Mohr, Freiburg.

von Plato, J. (1983). The method of arbitrary functions. *British Journal for the Philosophy of Science* 34:37–47.

Wallace, A. R. (1871). On the tendency of varieties to depart indefinitely from the original type. In *Contributions to the Theory of Natural Selection*, chap. 2, pp. 26–44. Second edition. Macmillan, London.

White, R. (2005). Epistemic permissiveness. *Philosophical Perspectives* 19:445–459.

————. (2010). Evidential symmetry and mushy credence. *Oxford Studies in Epistemology* 3.

Williamson, J. (2010). *In Defence of Objective Bayesianism*. Oxford University Press, Oxford.

Xu, F. and S. Denison. (2009). Statistical inference and sensitivity to sampling in 11-month-old infants. *Cognition* 112:97–104.

Xu, F. and V. Garcia. (2008). Intuitive statistics by 8-month-old infants. *Proceedings of the National Academy of Sciences* 105:5012–5015.

Zhang, K. (1990). Uniform distribution of initial states: The physical basis of probability. *Physical Review A* 41:1893–1900.

INDEX

Abrams, Marshall, 232n1
alsobability, 202
approach circle, 18
approach point, 18, 21
  stochastic independence of, 20
army blanket roll. *See under* die roll

babies. *See* children
balls, bouncing. *See* bouncing balls
barnacle
  Poli's stellate, 185, 188
  sexually dimorphic, 188, 204
Bayes, Thomas, 165–166, 168
Bayesianism, 31
  objective, 34
Bernoulli distribution, 38, 231n1
  learning empirically, 116
Bernoulli, Daniel, 240n11
Bertrand's chord problem, 30–31, 35
Bertrand's paradox, 30–31, 34, 186,
  221
  equidynamic solution, 69, 115, 207
Bertrand, Joseph, 30, 31
Bessel, Friedrich, 169, 170
Boltzmann, Ludwig, 8, 14, 124, 223
Boscovich, Roger, 164
bouncing balls, 206, 215
  adults' reasoning about, 46
  children's reasoning about, 43–45,
    108, 110–112
  equidynamics of, 93–112, 119–120,
    170–171

model of gas, 18
nonspherical, 23, 108, 119–120,
  121, 230n6
Brahe, Tycho, 162
braking pads. *See under* wheel of
  fortune
brief shaking. *See under* die roll; urn
  drawing
Browne, Janet, 130
Brush, Stephen, 229n1, 230n7

canceling out of errors, 162
Carey, Susan, 47, 92
Carnap, Rudolf, 32
central limit theorem, 20, 167
chaos. *See* sensitivity to initial conditions
children
  concept of probability, *see* physical
    probability, concept of
  equidynamic abilities, *see under*
    bouncing balls; urn drawing
  looking time studies, 39
Clark, Peter, 232n1
classical probability, 28–31, 224. *See
  also* indifference
  in children, 231n5
  dual nature, 29, 31–32, 47
coarse-graining, 61, 195, 199. *See also
  under* independence
  biology, 141, 156, 238n10
coin toss, 63, 152, 189
  bouncing phase, 71–74, 118, 206

complexity of equidynamics. *See under* equidynamic reasoning
concept of physical probability. *See under* physical probability
corollary to uniformity rule. *See under* uniformity rule
Cournot, Antoine-Augustin, 32, 47
Croson, Rachel, 40
curve-fitting, 163–164. *See also* least squares
Czuber, Emanuel, 242n1

Darwin's argument for evolution by natural selection, 129–134, 142–147, 149–153
  ecological stability essential, 150–152
  as equidynamic, 133, 144–147, 152–153
  gradualism, 149, 151
  parallel with Wallace, 237n7
  rhetorical force, 131, 152
  rhetorical force explained, 147, 153
  as statistical, 129–131, 143, 149
Darwin, Charles, 223. *See also* Darwin's argument for evolution by natural selection
  practical probabilistic expertise, 130
Daston, Lorraine, 29, 169
Denison, Stephanie, 38, 39, 41, 42
Diaconis, Persi, 233n9
die roll, 35, 74–78, 81–82
  adult reasoning about, 27–28
  alligator wall (craps table), 75, 234n4
  army blanket roll, 75, 83–84, 119, 137, 234n3
  bouncing phase, 75, 82, 115, 236n1
  brief shaking, 115, 234n8
  compared to urn drawing, 87–88
  container shape, 119
  dodecahedral, 27–28, 200–201
  equidynamics of, 33, 82–86, 113–116, 131–132
  initial conditions, 193, 202–203
  loaded, 115–116, 118, 130–131, 216

markings irrelevant to probability, 29, 35
  noise, 199
  setting, 84, 115, 207, 234n4, 234n5
  statistical investigation of, 75, 207, 234n5, 242n1
  three phases, 75, 207
Doors of Destiny, 27, 32, 33
Dupoux, Emmanuel, 92

economics, 180
Edwards, Paul, 174, 240n14
elementary errors, hypothesis of, 167–168
Ellis, R. L., 16–17, 30, 31, 221
embedding. *See* microconstancy, generated by embedding
Endler, John, 154, 155, 158
Engel, Eduardo, 71, 232n3
enion probability analysis, 180, 241n17
epistemic probability, 31–32, 33
  in development of statistics, 168–169
  use in single-case application of physical probability, 231n4
equidynamic reasoning, 3, 36, 226. *See also* equilibrium rule; independence rule; majority rule; microdynamic rule; microequiprobability rule; stirring rule; uniformity rule
  in children, *see under* bouncing balls; urn drawing
  complexity of, 118–120
  defeasible, 67, 138, 188–189, 207
  dynamic versus initial condition rules, 205
  dynamic versus nondynamic implementation, 90–92, 110–111
  evidence for, 208
  innate, 47–48, 67, 220, 224–225
  invisibility of, 130–132, 153, 179, 225–226
  not a priori, 35, 54, 154, 157–158, 170, 215–216
  practical utility of, 217–219
  precedence among rules of, 190, 191, 198

provisionality of rules, 208
reliability, 97, 197–198, 207–208,
    see also individual rules
short-term versus long-term rules,
    93–94, 100, 118–119
standard variables only, 69, 96,
    186–187, 207
universality, 45–47, 49, 207
equilibrium rule, 100–105, 112,
    213–214, 226, 235n2
    atmosphere, 171
    die, 114, 115
    gas, 121, 122
    organisms, 144–145, 218
    random walk model, 103–105, 106,
        114–115, 117, 145, 213
    relaxation, 101, 105, 213–214
    relaxation time duration, 103, 104,
        110, 114–115, 117, 118–119
    stochastic variable, 101–102, 213,
        226
    stochastic variable candidates, 114,
        117, 121, 144–145
    urn, 117
equilibrium rule package, 113, 206,
    226
    complexity of, 118
    die, 113, 115
    gas, 121
    natural selection, 152
    urn, 116, 117
ergodic theorem for Markov chains,
    104–105
ergodicity, 232n8, 235n11
error curve
    Gaussian, 165, 166, 169
    Gaussian, derivation of, 15–16,
        166–168
    independence of individual errors,
        162, 163, 168, 171
    Laplacean, 165, 166
    larger errors rarer, 162, 169
    properties equidynamically inferred,
        170–172
    properties statistically inferred, 169,
        240n12
    symmetry, 162, 165–166, 168–171
    triangular, 165

evolution function, 56, 226
    noise, 193, 199
evolutionary theory. *See also* Darwin's
    argument; natural selection
    equidynamic model-building,
        146–147, 155–159
    explanation of equidynamics, *see*
        equidynamic reasoning, practical
        utility of
    neutral models, 239n4
    optimization models, 154, 155
    as statistical, 127, 128
exogenous initial condition distribution.
    See under initial condition
    probability distribution
explanation
    of equidynamics, 217–219
    of frequencies, 54, 202–203

Feynman, Richard, 1
flowers and pollinators, 142–147, 154,
    155, 156–157
Fodor, Jerry, 220
Franklin, James, 48, 225, 231n3
freeze strategy (predator avoidance),
    134–138, 218

Galápagos finches, 151, 224
Galileo Galilei, 162, 169
Garber, Elizabeth, 17
Garcia, Vashti, 38
gas. *See* kinetic theory; Maxwell's
    derivation of velocity distribution;
    statistical physics; velocity
    distribution
Gauss, Carl Friedrich, 166–167, 168,
    240n6
Gaussian distribution
    of components of molecular velocity,
        8, 11–12
    of measurement errors, *see under*
        error curve
gerrymandered variable, 186, 196, 207
Gibbs, Josiah Willard, 8, 124, 223
Giere, Ronald, 230n4, 231n4
Gigerenzer, Gerd, 237n8
Gillespie, John, 239n1
Glymour, Clark, 243n1

Glynn, Luke, 232n1
Godfrey-Smith, Peter, 236n2
Grant, Peter and Rosemary, 151,
    223–224
gravitational field, gas in, 14, 107
Greggers, Uwe, 238n15
gun, firing of. *See* rifle, firing of
Gyenis, Balázs, 230n7

Hacking, Ian, 29, 47, 224, 225, 242n2
Hadley, George, 173
Hagen, Gotthilf, 167–168
Hald, Anders, 239n1, 239n2, 240n7
Halley, Edmond, 173
Helmholtz, Hermann von, 229n1
Hempel, Carl, 243n1
Herschel, John, 15–17, 30, 161, 169
history of science, methodology of,
    224–226
Hodge, M. J. S., 130, 237n4
Holland, John, 216
Hopf, Eberhard, 53, 232n3
Howson, Colin, 230n4

IAS computer, 174
independence rule, 171, 210
independence, stochastic
    causal independence as sufficient
        condition for, *see* independence
        rule
    children's reasoning about, 40–41
    of coarse-grained variables, 98–99,
        105
    concept, 48
    joint microequiprobability sufficient
        for, 60–61, 195, 235n1
    of measurement errors, *see under*
        error curve
    of microposition assumed by
        Maxwell, 20, 25, 26
    part of microequiprobability, 26
    required in stochastic variable, 102
    rough, 40
    of small-scale disturbances, 178
    wheel of fortune, 54, 59–61
    yielded by equidynamic rules, 83, 95,
        101, 106
indifference, principle of, 28–29

in development of statistics,
    168–169
dual nature, 29, 32–34, 35
failure to bifurcate, 35–36
Herschel's implicit use, 16–17, 169
objections to, 29–31, 34–35, 221
infants. *See* children
inflationary dynamics, 81, 226. *See also*
    sensitivity to initial conditions
initial condition probability distribution
    assumed to exist, 55
    doing without, 200–204
    epistemically exogenous, 185
    ur-conditions, 192–193
innateness, 231n3
    of equidynamics, *see under* equidyn-
        amic reasoning
instance confirmation, 221–224

Jaynes, E. T., 30–31, 34, 35, 36–37
Jeffreys, Harold, 34, 35, 36–37
Johnson-Laird, Philip, 45

Kahneman, Daniel, 215
Keegan, John, 160
Keller, Joseph, 63
Kelvin, Lord, 229n1
Kettlewell, Bernard, 223–224
Keynes, John Maynard, 30, 230n4
kinetic theory, 22, 171, 177. *See also*
    Maxwell's derivation of velocity
    distribution; statistical physics;
    velocity distribution
    hard sphere model, *see* bouncing
        balls
    reception, 8–9, 229n1
Kolmogorov Arnold Moser (KAM)
    theorem, 235n4
Kries, Johannes von, 30, 232n3
Krüger, Lorenz, 48

Labby, Zacariah, 242n1
Lambert, J. H., 240n10
Laplace's demon, 132–133, 156
Laplace, Pierre Simon
    classical probability and indiffer-
        ence, 28–29, 30, 36–37, 45, 215,
        231n6

error curve, 166, 167, 169, 240n11
  statistical methods, 163, 164
law of large numbers. *See also*
    statisticalism
  ergodic theorem for Markov chains,
    104–105
  evolutionary biology, 129–130, 149,
    237n7
  washing out, 20, 21–22, 46, 93,
    103–104, 157, *see also* equilib-
    rium rule, random walk model
least squares, method of, 15, 164
  justification of, 164–168
Legendre, Adrien-Marie, 164
Leibniz, Gottfried Wilhelm, 215
levels of information. *See*
    coarse-graining
Lewis, David, 32
Lewis, John, 174
Loewer, Barry, 232n1
looking time in children, 39
Lotka, Alfred, 179, 180–181
Lotka-Volterra models. *See* population
    ecology
Lynch, Peter, 173

macroperiodicity. *See*
    microequiprobability
magnetic claw. *See under* urn drawing
majority rule, 136–139, 190, 211, 226
Marinoff, Louis, 30, 230n2
Maxwell's derivation of velocity
    distribution, 7–8
  as a priori, 9, 10–11, 14–15, 21, 35,
    123
  empirical confirmation, 8, 9
  Herschel's influence, 15–17, 225
  as lucky, 15, 17
  mathematical proof, 11–12
  official and unofficial distinguished,
    21, 24, 120
  official as afterthought, 24–25,
    243n3
  official, structure of, 9–10, 13–14
  propositions i–iii, 17–19, 100–101
  proposition iv, 13–14, 17–18,
    19–20, 24, 27
  proposition vi, 24, 124

proposition xxiii, 23–24, 124
  relation of propositions i–iii to
    proposition iv, 20–24, 120–121
  rhetorical force, 13, 21, 27
  rhetorical force explained, 24, 36,
    48–49, 121, 122–123
  role of equidynamics in, 36, 48,
    121–122
  second derivation (1867), 24, 124
  unofficial, structure of, 21–24,
    120–122
Maxwell, James Clerk, 8, 123–124,
    223. *See also* Maxwell's derivation
    of velocity distribution
Maxwell-Boltzmann distribution. *See*
    Maxwell's derivation of velocity
    distribution; velocity distribution
Mayer, Tobias, 163
measure-relativity, 68
measurement error. *See also* error
    curve
  empirical confirmation of Gaussian
    distribution, 169
  hunger's effect on, 172
  physics of, 170
Mehler, Jacques, 92
Menzel, Randolf, 238n15
meteorology. *See* troposphere; weather
    forecasting
method of arbitrary functions, 232n3
"micro", standard for, 69–70, 96, 198,
    209
  2% rule, 187
  physiological, 191, 210
  strength, 99
micro-orientation, 144
microconstancy, 56–57, 226
  generated by embedding, 81, 82
  generated by stirring, 62
  relation to random walk, 104–105
  relativity, 68–70
microdirection, 114, 226
microdynamic rule, 94–100, 112,
    212–213, 215, 226
  atmosphere, 171
  bouncing balls, 102, 108
  die, 114, 115
  gas, 121, 122

microdynamic rule *(continued)*
   microequiprobability assumption
      implicit, 189
   organisms, 139–142, 144
   source of stochastic variables, 102
   trumps microequiprobability rule,
      190, 198
   urn, 117
microequiprobability, 26–27, 58, 226
   of actual initial conditions, 201
   amplification, 99–100, 195–196
   biological, 139–142
   knowledge of, 65–67, *see also*
      microdynamic rule; microequi-
      probability rule
   nonprobabilistic tendency to,
      202–203
   perturbation explanation, 192–199,
      204
   physiological explanation,
      190–191
   relativity, 68–70
microequiprobability rule, 185–190,
      197, 198, 209, 226
   atmosphere, 177, 178
   mosaic, 192, 203
   noise, 200
   organisms, 139, 218
   physiological, 160, 171, 191–192,
      209–210, 218, 226
   wheel of fortune, 67
   without initial condition probability,
      204
microlinearity, 94, 95–96, 226
   nonmicrolinearity, 96, 197
   relativity, 96
   role in creating microequiprobabil-
      ity, 97–98, 197
   transformations preserve microequi-
      probability, 197
microposition, 95, 226
model-building. *See* economics;
      evolutionary theory; physical
      intuition; population ecology;
      troposphere; war
molecular velocity distribution. *See*
      Maxwell's derivation of velocity
      distribution; velocity distribution

mosaic microequiprobability rule. *See*
      under microequiprobability rule
motley wheel. *See under* wheel of
      fortune
motor control, 66, 191

natural selection. *See also* Darwin's
      argument; evolutionary theory
   backfiring trait, 132, 135, 155
   complications, 139, 154, 237n9,
      239n1
   conditions for, 127–128
   direct confirmation of, 223–224
   ecological stability essential,
      150–152
   fitness, 128, 134, 139, 236n2
   as stochastic, 128, 156
noise, rationale for ignoring, 199–200
North, Jill, 32, 233n11, 241n5
Norton, John, 34
Novack, Greg, 34

Oster, George, 155

Pearson, Karl, 207
perturbation explanation for microequi-
      probability. *See under*
      microequiprobability
Phillips, Norman. *See under*
      troposphere
physical intuition, 1–3, 83, 134, 205.
      See also equidynamic reasoning;
      probabilistic dynamics; propor-
      tional dynamics
   model-building, 2, 155
   not evidence, 222–223
physical probability, 31, 32, 54
   alsobability, 202
   concept of, 47–48, 224–225
   metaphysics, 55, 201–202
   possibly transient or nonexistent in
      ecosystems, 150
   single-case application, 39–40, 48
physiological microequiprobability rule.
      See under microequiprobability
      rule
Plackett, R. L., 162
Plato, Jan von, 232n3

Poincaré, Henri, 53, 232n3, 242n10
Polanyi, Michael, 1–2
population ecology, 179–180, 223
   stable behavior, 150–153, 179
Porter, Theodore, 230n5
probabilistic dynamics (faculty), 104, 106, 206, 213–214, 226
   bouncing balls, 108–110
   complexity of, 119
   die, 114–115
   gas, 121–122
   Monte Carlo implementation, 111–112
   organisms, 218
   and symmetry, 215
   urn, 117
probabilistic independence. *See* independence
probability. *See also* classical probability; physical probability; epistemic probability
   adult mishandling of, 46, 215–216
   dual nature, *see under* classical probability
   inference about, *see* equidynamic reasoning; indifference
probability matching, 238n15
proportional dynamics (faculty), 73–74, 159, 190, 205–206, 226
   atmosphere, 178
   bouncing balls, 103, 111
   complexity of, 119
   die, 114, 116
   organisms, 133, 137–138, 141–142, 218
   shooting, 161, 218
   and symmetry, 215
   urn, 117–118

Quetelet, Adolphe, 230n5

random variable, 68
random walk. *See* shaking; equilibrium rule
Reichenbach, Hans, 53, 231n4, 232n1, 232n3
relaxation. *See under* equilibrium rule

Richardson, Lewis Fry, 173, 181
rifle, firing of, 15, 161
Rogers, William, 172
rogue variable. *See* gerrymandered variable
Rohrlich, Fritz, 2
roulette wheel, 206

saber-toothed tiger, 27, 218
Salmon, Wesley, 232n1
Samuelson, Paul, 180
Schmidt, Richard, 191
sensitivity to initial conditions, 232n5. *See also* sensitivity-symmetry rule
   bouncing ball reasoning, 46
   bouncing balls, 95, 102, 108
   die, 76, 82
   equilibrium rule, 101, 103
   majority rule, 136, 137
   microconstancy, 62
   microdynamic rule, 94
   more important than symmetry, 215
   motley wheel, 81
   organisms, 140–142, 145
   role in creating microequiprobability, 97–98, 195
   shooting, 160–161
   undermines ecological stability, 152
   urn, 87, 91
   weather, 177, 240n15
sensitivity-symmetry rule (putative), 64–65, 83–84
Shackel, Nicholas, 230n2
shaking
   brief, *see under* die roll; urn drawing
   parameter, 85, 86, 90, 113, 226
   process, 84–86, 90, 104, 206, 226
Sheynin, O. B., 130, 237n6, 240n10, 240n13
Simon, Herbert, 180–181
Simpson, Thomas, 165–166
single-case application of physical probability. *See under* physical probability
Sklar, Lawrence, 235n4

smoothness
   component of microdynamic rule,
     94
   of dynamics, *see* microlinearity
   of probability distribution, *see*
     microequiprobability
Somme, Battle of, 160
Sperber, Dan, 92
standard variable, 69, 226
   in equidynamics, *see under* equidyn-
     amic reasoning
   tendency to microequiprobability,
     69, 196–197
statistical mechanics. *See* statistical
   physics
statistical methods. *See* curve-fitting;
   take the average; least squares
statistical physics, 3–7. *See also* kinetic
   theory
   development of, 124
   equipartition approach, 124
   formalist ideology, 223
statisticalism (errors cancel out), 162
Stern, Otto, 9
Stigler, Stephen, 161–162, 168, 169,
   239n1, 240n11
stirring
   parameter, 62, 85, 226
   process, 62–63, 215, 226, 235n11
stirring rule, 65–66, 67, 210–211,
   226
   microequiprobability assumption
     implicit, 67, 189
stochastic independence. *See*
   independence
stochastic variable. *See under* equilibrium rule
strike ratio, 59, 226
subjective probability. *See* epistemic
   probability
Sundali, James, 40
symmetry. *See also* indifference;
   sensitivity-symmetry rule
   role in equidynamics, 214–215

take the average, 161–162
   aggregate the equations, 162–163
   justification, 162, 163, 165–166

take the mean, 162, 167
take the median, 239n2
take the mean. *See* take the average
Téglás, Ernő, 43, 111
telescopic measurement, 15, 161,
   170–172
Tenenbaum, Joshua, 111
Thomson, William, 229n1
three-legged table, 1
Todorov, Emanuel, 191
Tolman, Richard, 223
Tolstoy, Leo, 179
trench warfare. *See under* war
troposphere
   equidynamic reasoning about,
     177–178
   freak disturbance, 175–177
   Phillips model, 173–175
   simulation, 173–177
   structure, 172–173
Truesdell, Clifford, 8, 230n7
Tversky, Amos, 215
twinkling, 170–171

uniformity rule, 105–108, 214, 226
   atmosphere, 171
   bouncing balls, 108
   corollary, 106–107, 108, 214
   die, 114, 115
   gas, 121, 236n3
   organisms, 145–146, 218
   urn, 117
Urbach, Peter, 230n4
urn drawing, 86–89, 110, 111, 133
   adults' reasoning about, 45
   biological analogy to, 144, 218
   brief shaking, 88–90, 117
   children's reasoning about, 38–43,
     90–92
   equidynamics of, 89–90, 116–118
   magnetic claw, 117–118, 119, 205,
     216
   nonrandom sampling, 42, 91
   sticky balls, 41–42, 91, 92
   target zone, 87, 88, 89, 116
   two phases, 86
   with and without replacement,
     87

van Fraassen, Bas, 30
velocity distribution (of molecules in a gas), 7–8. *See also* Maxwell's derivation of velocity distribution
Venn, John, 221
vernier, 170, 171
visitation rate, two kinds, 85–86. *See also* shaking process; stirring process
Viswanathan, Gandhimohan, 238n15
Volterra, Vito, 179
von Kries, Johannes. *See* Kries, Johannes von
von Neumann, John, 173
von Plato, Jan. *See* Plato, Jan von

Wallace, Alfred Russel, 134, 237n7
war
    modeling outbreaks, 181
    trench warfare, 160–161, 173, 181

washing out. *See under* law of large numbers
Waterston, J. J., 124
weather forecasting, 173, 176
    chaos, 240n15
Weldon, Raphael, 207
wheel of fortune, 53–61, 62
    braking pads, 54, 118, 130–131
    equidynamics of, 64–65
    motley, 78–81
White, Roger, 34, 230n3
Williamson, Jon, 231n5
Wilson, E. O., 155
Wolf, Rudolf, 242n1
wolves and deer, 129–134, 138–139

Xu, Fei, 38, 39, 41, 42

Zhang, Kechen, 71, 232n3